Code Cool Stuff With Python

By Doug Purcell

In the book are links to several 3rd party sites. This does not mean that the author endorses the site or its products. No liability is assumed from the installation of any software from these sites. It's recommended to have the latest antivirus and spyware software installed on your machine to combat the latest security threats.

Paperback ISBN: 978-0-9973262-7-7

Connect with me!

I like meeting new people, especially dudes and dudettes that dig my work. The social media channel I'm most active on is LinkedIn so feel free to send me a connection request along with a quick message as to how your python programming journey is coming:
https://www.linkedin.com/in/doug-purcell

Table of Contents

A Merry Overview of the Python Programming Language

Learning the basics of any language isn't necessary the funnest parts, but it's definitely a necessary evil. I've condensed the basics of python into a single chapter so that we can start building more interesting programs in the subsequent chapters. Learn how to install python on your machine, get acquainted with the free PyCharm editor, learn primary data types, master core data structures, build reusable code with functions, study object oriented programming, and learn how to construct try/except statements.

Getting python installed on your machine

Just like python is renowned for its simplicity, the same can be said about its installation procedure. Below are the steps on how to quickly install python on all of the major operating systems. If you already have python installed then you can fast-forward to the *Setting up PyCharm* section.

Install Python on Windows

There's a high probability that if you're using a Windows powered machine that Python won't be there by default. To discover if Python is indeed installed you can open up command prompt by typing *cmd* in the search bar followed by the word *python*. If it's there then the python interpreter will run; otherwise you'll get an annoying message like the following:

```
'python' is not recognized as an internal
or external command, operable program or
batch file
```

This tells you the sad news that Python is not installed and that you have to set it up. Follow the steps below to make that happen:

Step one: Download the latest version of python on your machine: https://www.python.org/downloads

Step two: Start the Windows installer that matches your system. If you click "Install Now" then Python is installed in the "user" directory, but if you change its location then just remember where it's installed.

Step three: You'll have an option to add Python to PATH which is where the computer searches for Python when you type it via command prompt. If you check this box then Python will be available via this option, if not then an error will occur.

Therefore, it's a good idea to go ahead and check this option so that you can type in python commands via command prompt. If you installed Python without selecting this option then no biggie as you can manually add the path to your system. Below are the steps:

- In the Windows menu search for *advanced system settings*.
- In the window that displays click ENVIRONMENT VARIABLES.
- In the next window, find and select the user variable called path and click EDIT.
- Scroll to the end of the value and add a semicolon (;) followed by the location of PYTHON.EXE. If you didn't change the default installation location it should be located in your user directory.
- Click OK to save the settings.

If you don't know the location of python.exe then don't panic, just search for PYTHON.EXE in the Windows menu. Once located, right click the file, select properties, and view the Location. Right click to copy the full path and then paste it at the end of the Path user variable.

If you don't have one then click the new button, add a variable named Path, and then add the value which is the location or "path" of the `python.exe` file. Once done type "python" into the terminal to ensure that everything was set up properly and that it runs.

Install Python on macOS

Like Linux, Python is already installed on a variety of OS X systems. You can confirm that Python is installed by going to: APPLICATIONS → UTILITIES → TERMINAL. Next, type the following into the terminal:

```
$ python --version
```

The command will output the version of Python which is:

```
Python 3.7.3
```

One way to install the latest-and-greatest version of python on macOS is to use the appropriate macOS installer that matches your system:
https://www.python.org/downloads

Installing Python on Linux (Ubuntu 18.04)

To see if Python is installed on your machine open up the terminal and type in the following:

```
$ python
```

You can fire-up the terminal by using the keyboard shortcut: CTR + ALT + T.

The output should look something like the following:

```
Python 3.7.3 |Anaconda, Inc.| (default, Apr 29
2018, 16:14:56)
[GCC 7.2.0] on linux
Type "help", "copyright", "credits" or
"license" for more information.
```

If you got something like this then WOOT-WOOT, Python 3.7.3 is installed on your machine. If Python 2.7 or later is installed then it's OK, you don't need to uninstall it, you just need to get Python 3.0+ running. Luckily this process is super easy with Ubuntu:

- Step one: Open up the terminal by typing: `ctr + alt + t`
- Step two: `sudo apt-get update`
- Step three: `sudo apt-get install python3.7.3`

The word *sudo* is abbreviation for "super user do" and it allows programs to be executed as a super user, aka the root user. The apt command means Advanced Package

Tool, which is a package manager for Debian based operating systems like Ubuntu. The apt-get command is the APT package handling utility.

A short term alternative is to use an online python interpreter. Here's some of the following:

- Online GDB:
  https://www.onlinegdb.com/online_python_interpreter

- Repl.it: https://repl.it/languages/python3

- Another online python interpreter:
  http://mathcs.holycross.edu/~kwalsh/python

A Quick Tutorial to Learning PyCharm

There are many choices of integrated development environments (IDEs) in python that you can choose from such as sublime, IDLE, Vim, Wing, Atom, Spyder, and PyCharm. There's like way too many for me to keep track of so you can always check out Wikipedia: http://bit.ly/2Px8AYs

If you have python installed, an editor you're comfortable with, and ran the standard *Hello World* program then you can jump to *The One Day Python Bootcamp* section in this chapter. If you don't already have an editor or IDE then I'll recommend PyCharm which has the largest mind share in the python community.

Before you download and run PyCharm you need to have the python interpreter installed. There are three flavors to

PyCharm which are the community, education, and professional editions. The community edition is open source and compared to the commercial version (professional), it comes equipped with fewer features. However, when you're starting out programming the community edition will suffice. Below are the instructions on the various operating systems:

Windows:

- Download the PyCharm installer, run the executable file, and follow the installer steps. Here's the instructions on the PyCharm website: http://bit.ly/2q5Y2Vx

- **MacOs:** Download the PyCharm disk image and mount and drag the image to the Applications folder: http://bit.ly/331bclr

- **Linux**: If you have Ubuntu 16.04 you can install PyCharm through the command line using the snappy package manager:

    - ```
      $ sudo snap apt-get install
      pycharm-community
      ```

Hello World with PyCharm

Once you've installed PyCharm the next step is to run the proverbial *Hello World* program. Here's the step-by-step procedure on how to do this.
      ```

Create a Brand Spanking New Project

Create a new project by doing the following: *File → New Project*

A project is an organizational unit in PyCharm. Name the project *SampleProjects;* here's a screenshot of what the setup should look like thus far:

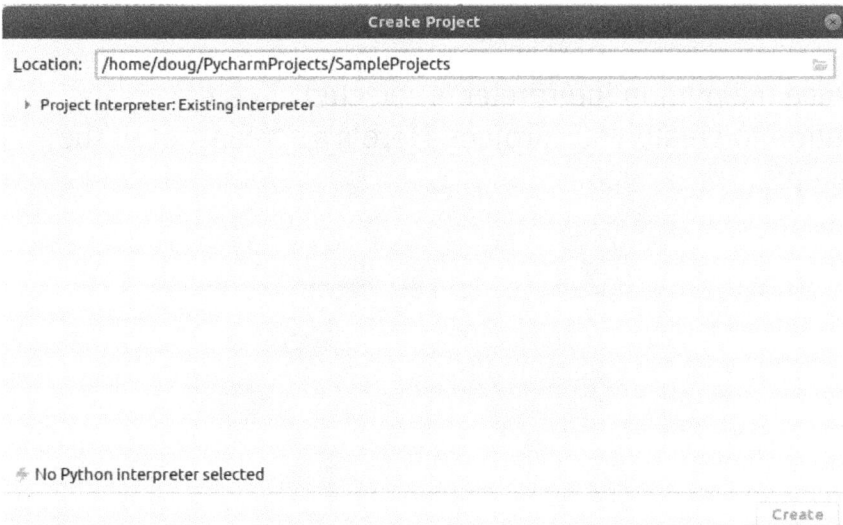

Create Project	⊗

Location: /home/doug/PycharmProjects/SampleProjects

▶ Project Interpreter: Existing interpreter

⚡ No Python interpreter selected

Create

Image: Create SampleProjects.

Next, select the python interpreter that you want to use. Click the arrow that's next to *Project Interpreter:Existing Interpreter,* and then select a python 3.6 or up. Below is an example of how things should look:

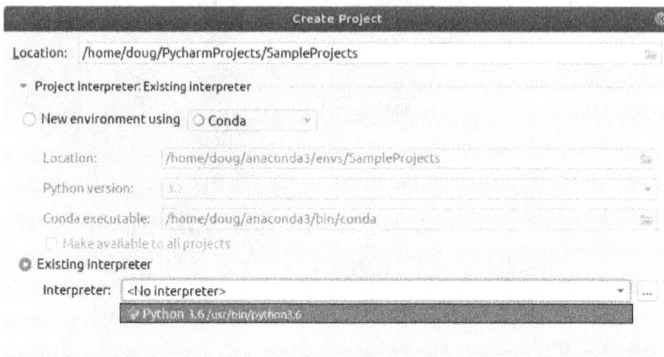

Image: Adding Interpreter.

Once the python interpreter is selected click the create button to create your project. Here's a screenshot of the setup:

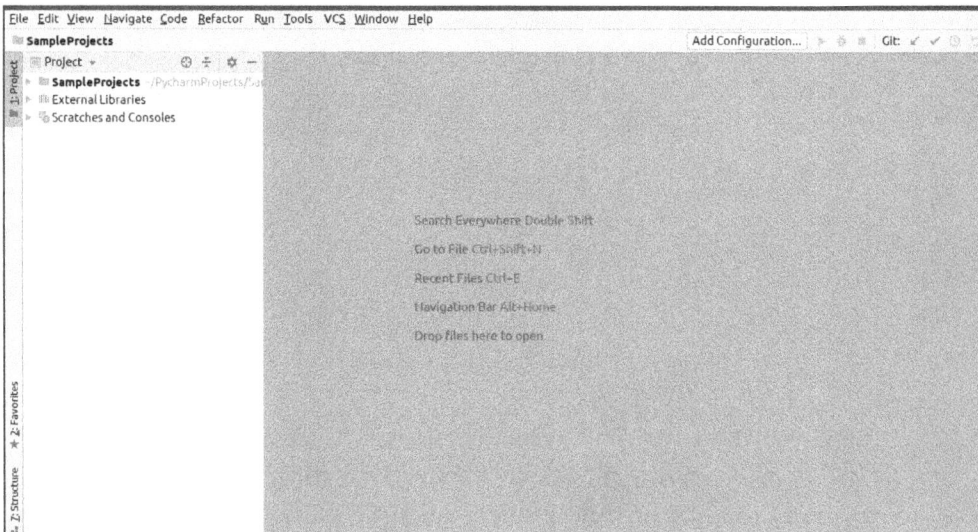

Image: PyCharm project setup.

Create a fresh python file

At the main menu on PyCharm, or the portion where you see the various menus such as File, Edit, and View, do the following: **File → New File → python file**

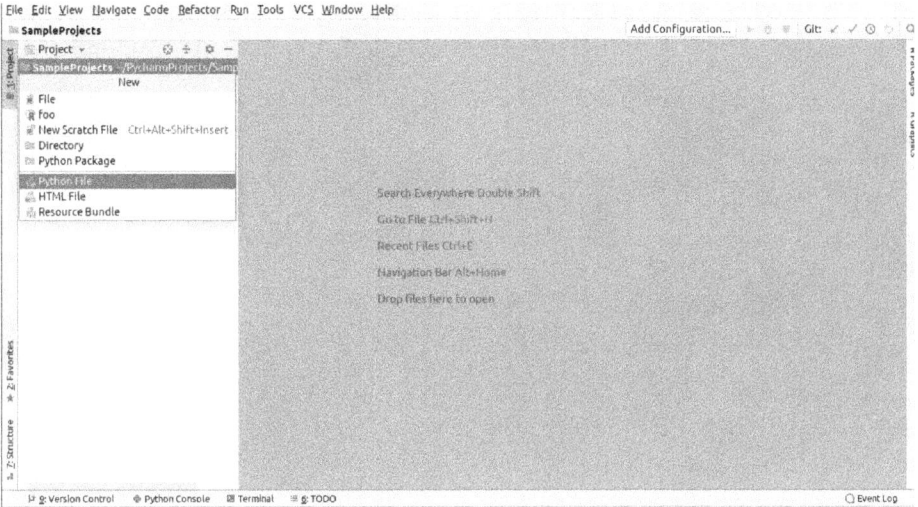

Image: Create a fresh python file.

A text box should appear which looks like the following:

Image: New Python file.

Enter in the name *HelloWorld* and select the OK button. Once done here's how your project setup should look:

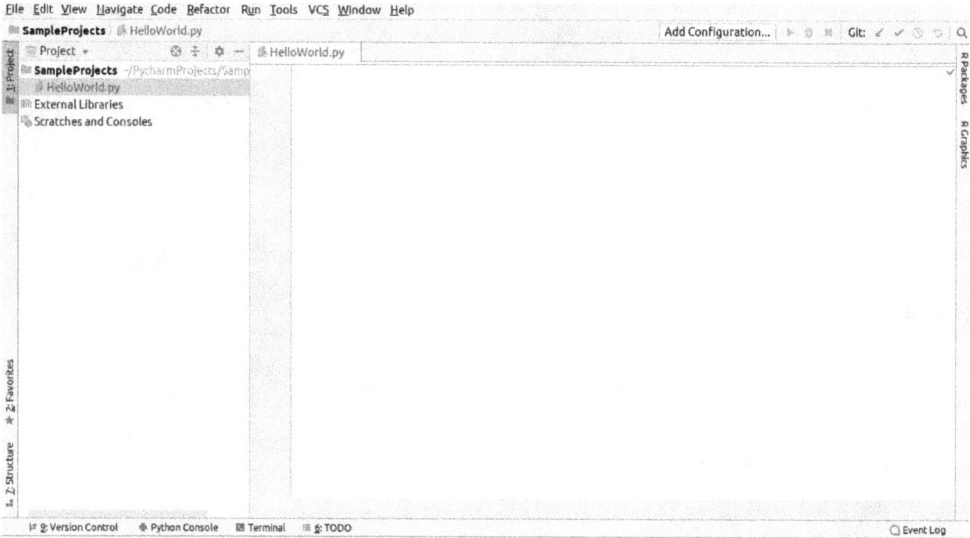

Image: HelloWorld.py in PyCharm.

The blank white space is known as the editor. That's where you'll spend most of your time hacking away. The left hand side is known as the project manager, that's where you can see the organization of files in a project.

Add code and run the file

Copy the following code into the editor:

```
print('Hello World!!!')
```

To run the file click the following on the main menu: `run` → `run`

A dialog box should appear which looks like the following:

Image: Run HelloWorld

Select the *HelloWorld* program. Once the program has executed you should see the text: *Hello World!!!*

If all went well then congrats, you've ran your first python program. Now, there's already a plethora of free curated information about PyCharm online. Here are a couple of places to whet your appetite:

- Quick Start Guide: http://bit.ly/31XottL

- PyCharm blog:
  https://blog.jetbrains.com/pycharm

- Learn keyboard shortcuts for editing, navigating, refactoring, and debugging:
  http://bit.ly/2PuCMU9

The One Day Python Boot camp

You can learn the gist of python in a day; I have no doubt about that. It will be a somewhat superficial level, but it's a start nonetheless that you can build on top of overtime. If you want to gain a deeper understanding of the basics of python then I'll wholeheartedly recommend the best book for learning python basics which is *Become a Python Developer:* https://amzn.to/2Snerhn

It's written by the amazing author Doug Purcell... oh shoot, my cover is blown.

Variables

A variable in python is similar to a variable in mathematics. It's something that has a changeable state. Examples of variables in python are shown below:

```
a = 10

b = 1.598

c = .1987
```

```
d = 100.579
```

Printing output

The above code snippet simply stores the variables in the computer's memory. This means that the data is there but you as the user can't see it. In order to view the data you need to use the `print` function to display the output as shown below:

```
print(a)
print(b)
print(c)
print(d)
```

Swapping variables

A useful tip to know in python is that to swap variables you can do that in a single statement.

```
x = 5
y = 10
z = 30
x, y, z = z, x, y
print(x)
print(y)
print(z)
```

The output will be:

```
30

5

10
```

Variable Naming Tips

For details on how to properly name variables in python refer to the python enhancement proposals also known as *PEP 8*.

Here are some of the highlights of PEP 8:

- Variable names can have letters, numbers, and underscores.

- Can't use a reserved word like print.

- Be as descriptive as possible with your variable names. This reduces ambiguity and helps make your code more maintainable when other developers follow in after you.

- Python IS case sensitive so apple is not the same as Apple.

- Put constants or variables that value is fixed in all CAPS. I.e., `DAYS_OF_WEEK = 7`

For a more comprehensive overview refer to PEP 8:
https://www.python.org/dev/peps/pep-0008

Python Math Operators

The word computation has computer in it, which gives a hint to one of the uses of computers. Python like many sophisticated programming languages can be used as a *souped up calculator*. All of the standard features that are available on scientific computers can easily be emulated with the help of built in operations and modules in python. Let's look at some of the math operators available in python: +, -, *, **, /, //, %

Most of these symbols you're probably already familiar with. Let's dig into some code to better understand this:

```
print(10 + 10)
print(50 - 10)
print(10 * 10)
print(20 ** 2)
print(9 / 5)
print(8 // 3)
print(11 % 5)
print(1e10)
```

Here's the output:

```
20
```

```
40
```

```
100

400

1.8

2

1

10000000000.0
```

The +, -, *, and / symbols behave as we expect. The asterisk means multiplication and forward slash means division. The double star, (**) means rise to the power. So, in this case 20 ** 2 means 20^2 or 400. The double forward slash (//) indicates the floor operator in mathematics. This means to divide the dividend by the divisor, and ignore the remainder. In this case this means 8 divided by 3, which is 2.66666666667, but the floor operator ignores all of the stuff following the decimal point (mantissa), so the answer is 2.

The % sign indicates modulus, so you divide the dividend by the divisor like you would with regular division except you take the remainder. Therefore you do 11 divided by 5 which is 2, and then take the remainder which is 1.

Adding additional functionality into your programs

If we use just the builtin math operators then we're severely limited in what we can do for our mathematical calculations. Luckily there's the math module which you

can check out here:
https://docs.python.org/3/library/math.html

It includes mathematical properties like logarithms and trigonometry. To use these functions in your program you need to use the import statement. Below is a quick example:

```
import math
print(math.log(1000000, 2))
print(math.sqrt(9))
print(math.cos(100) + math.sin(90) + math.tan(90))
print(math.pi**2 * math.e)
19.931568569324174
3.0
-0.23888487632000044
26.828366297560617
```

Strings

If you have ever sent an SMS text, used Facebook chat, or sent an email then you have used strings. A **string** is a sequence of characters wrapped in quotes; in python it could be single, double, or triple quotes. The single or double quotes can be used interchangeably. The triple quotes are typically used as doctrings, or comments inside methods, functions, or classes.

They're typically used when you need to include text that expands multiple lines as they can handle line breaks nicely. The best way to understand the difference between the various string types is to create a simple python program and experiment with them. Below is a quick overview of strings in python:

```python
city = 'Los Angeles'
# indexing: python is a zero based indexed language
print(city[0]) # L
print(city[3]) # empty space is a string!
print(city[4]) # capital A
print(city[-1]) # negative indices are permitted
# len() function: gets the length of the string
print(len(city)) # 11
print(city[len(city)-1]) # s
# concatenation: the combining of multiple strings
print('john ' + 'doe ' + 'public') # john doe public
# slicing: retrieves ranges of a string
print(city[0:3]) # Los
print(city[4:11]) # Angeles
print(city[::]) # Los Angeles
print(city[::2]) # LsAgls
print(city[::-1]) # selegnA soL
```

Boolean algebra

This is a branch of mathematics that was invented by English mathematician George Boole back in 1847. Even though it's over a century old its impact still permeates. It has been fundamental in the development of digital electronics, and is available in all modern day programming languages. Learning Boolean algebra for python means that you can apply that set of logic to a wide array of languages like Java, C++, Haskell, Erlang, or R.

In python what we need to worry about are the values of `True` or `False`, which can also be denoted by 1 or 0 respectively. The main operations that we'll discuss are "and" (conjunction), "or" (disjunction), as well as "negation." There's also the lesser used "xor" operator:

Below is a sample of how the truth table looks:

x	y	x and y	x or y	not x
0	0	0	0	1
1	0	0	1	0
0	1	0	1	1
1	1	1	1	0

Remember, 0 maps to `False`, and 1 map to `True`. Here's a shortcut to remembering this: *and* is always `False` or less you have two `True` operands, while *or* is always `True` or less you have two `False` operands. If you're confused about this then no worries, just commit the above table to memory. You'll need to remember it in order to understand conditional logic.

```
is_the_sky_blue = True
do_cats_bark = False

print(is_the_sky_blue) # True
print(do_cats_bark) # False
```

Remember, 0 and 1 could be used for **False** or **True**. Therefore you could use them interchangeably if you desire, even though **True** or **False** are more common.

The 'and' truth table

The 'and' operator evaluates to **False** in all situations EXCEPT when both operands are **False**.

```
print(True and True) # True
print(True and False) # False
print(False and False) # False
print(False and True) # False
```

The 'or' truth table

Or evaluates to True with at least one True operand.

```
print(True or True) # True
print(True or False) # True
print(False or False) # False
print(False or True) # True
```

The 'xor' truth table

Xor is a little tricky. It evaluates to True when the two operands are different.

```
print(True ^ True) # False
print(True ^ False) # True
print(False ^ True) # True
print(False ^ False) # False
```

Control Flow in Python

Once you understand Boolean algebra you can apply that newfound knowledge to control flow, or the order in which statements are executed. Python uses the `if`, `else`, and `elif` statements for this.

if/else statement

Here's an example of an `if/else` statement:

```
x, y, z = 5, 10, 15
if x < y and z > y:
    print(x)
else:
    print(y)
```

The `if` keyword is a reserved keyword in python, and the expression most be terminated by a colon. If the first expression is `True` then the statements inside the body are executed, if it's **False** then the branch under the `else` statement is executed.

elif statement

Below is an example of the `elif` statement in python:

```
from random import randint
# picks a random number in range 1...100
grade = randint(1, 100)
if grade >= 90 <= 100:
    print('A')
```

```
elif grade >= 80 <= 89:
    print('B')
elif grade >= 70 <= 79:
    print('C')
elif grade >= 60 <= 69:
    print('D')
else:
    print('F!')
```

Ternary Statement

This is a special type of operator that evaluates something based on a condition being True or False. The best way to understand it is to take a look at a simple code snippet:

```
mood = True
state = 'nice' if mood else 'not so nice'
print('state = {}'.format(state))
```

The following prints nice because if mood evaluates to True.

Comments

At this point you may have saw the hash symbol (#) followed by text. This is known as a *comment* in python and this portion of the code is ignored by the interpreter.

However, it's still very useful to include in your programs as it helps other programmers that may be messing around in your code to understand the logic.

 # This is a comment
 # This is a comment and will be ignored by the interpreter
 # I think you get the memo!

Data structures in python 3.7.3

There are four built-in data structures in python which are lists, tuples, dictionaries, and sets. Here's a quick rundown of each one:

lists are mutable collections of objects

Below is a demo of some of the features of a list in python:

```
evens = [0, 2, 4, 6, 8, 10]

# reverses the list

>>> evens.reverse()

[10, 8, 6, 4, 2, 0]

# adds an object to the list

>>> evens.append(100)

[10, 8, 6, 4, 2, 0, 100]

# merges another list with the list

>>> evens.extend([1, 3, 5, 7, 9])
```

```
[10, 8, 6, 4, 2, 0, 100, 1, 3, 5, 7, 9]

# pops an item from the list

>>> evens.pop()
```

Tuples

These are an immutable sequence. Unlike lists once you create a tuple they cannot be modified, and trying to do so will cause an error.

```
nums = (1, 3, 5, 7)


>>> print(nums)

...

(1, 3, 5, 7)
```

Dictionaries

These are key/value pairs or associative arrays in some languages. Here's a demo of dictionaries in python:

```
vowels = {'a': 0, 'e': 0, 'i': 0, 'o': 0, 'u': 0}

>>> vowels.items()

dict_items([('a', 0), ('e', 0), ('i', 0), ('o', 0),
('u', 0)])

>>> vowels.get('a')
```

```
0

>>> vowels.values()

dict_values([0, 0, 0, 0, 0])

>>> vowels.keys()

dict_keys(['a', 'e', 'i', 'o', 'u'])

>>> vowels.pop('e')

0

>>> vowels

{'a': 0, 'i': 0, 'o': 0, 'u': 0}
```

Sets

The set data structure stores unique items. Here's a demo of a set in python:

```
>>> letters = {'a', 'a', 'a', 'b', 'b', 'b'}

>>> letters.intersection({'b', 'c'})

{'b'}

>>> letters.pop()

'a'

>>> letters.add('c')

>>> letters.union({'a', 'e', 'i', 'o'})

{'i', 'c', 'e', 'b', 'o', 'a'}

>>> sorted(letters)

['b', 'c']
```

Iteration in Python

Iteration is the process in which computers do repetitive tasks. Humans despise repetition while computers are amazing at it. Humans can do repetitive tasks like summing all of the numbers from 1-100 manually (assuming no mathematical formulas are used), but these tasks are tedious and error prone. Computers can do number crunching like this in very quick times, like in a couple of freaking nanoseconds. The two main ways to do iteration in python is by using either the while or for loops.

WHILE LOOP

A `while` loop states that while a condition is `True` to execute the statements in the body.

```
# sets while loop starting at
i = 0
# condition
while i < 10:
    print('i = {}'.format(i, end=' '))
    i += 1 # increment i
```

When ran the above code prints 0 ... 9.

Here's another example:

```
# Sum numbers from 1...1000 in nanoseconds
i, sum = 0, 0
while i < 1000:
    i += 1
    sum += i


>>> print('The summation of 1...1000 = {}'.format(sum))

...

The summation of 1...1000 = 500500
```

for loop

This is another way to iterate in python. It can be used with the range function to iterate over a sequence of numbers or it can be used standalone to iterate over data structures like lists or sets. Below is a simple example of a for loop in python:

```
for x in range(10):
    print(x, end=' ')
```

The above prints the numbers 0 ... 9 on the same line separated by a space.

Fibonacci numbers

The following prints the 12th Fibonacci number:

```
x, y = 0, 1
for z in range(10):
    next = x + y
    x, y = y, next
print('12th fib number = {}'.format(next))
```

```
>>> print('12th fib number = {}'.format(next))

...

12th fib number = 89
```

Creating functions in python

A function is a set of inputs that map to a set of outputs. You can create your own functions in python by using the def keyword.

```
def scale_number(num, amount):
    return num * amount

>>> print(scale_number(10, 5))
...
50
```

Keyword arguments

```
def area_triangle(height=11, width=7.5):
```

```
    return 1/2 * (height * width)
```

```
>>> print(area_triangle())

...

41.25
```

```
>>> print(area_triangle(height=20, width=100))

... 1000.0
```

Accepting an arbitrary number of input

You can do this by appending an asterisk in front of the variable.

```
def multiply(*args, y=1):
for x in range(len(args)):
    y *= args[x]
return y
```

```
>>> print('multiply=', multiply(1, 2, 3, 4))

...

multiply= 24
```

Reading in an arbitrary number of keyword arguments

You can accomplish this by sticking two asterisks in front of the variable name.

```
def key_value(**kwargs):
for key, value in kwargs.items():
    print('{} {}'.format(key, value))
```

```
>>> key_value(a=5, b=10, c=15)

a 5

b 10

c 15
```

Classes and Objects

Object oriented programming is a style of programming that involves the heavy use of classes and objects. Classes are typically described as blueprints, while objects are described as the templates that are created from the classes. Below is a simple example of how to create a class in python:

```python
class Point:
    """Simple class in python. This is an example
    of a docstring, or a string that's used like a
    comment to document a segment of code."""

    def __init__(self, x, y):
        self.x = x
        self.y = y

    def get_x(self):
        return self.x

    def get_y(self):
        return self.y

    def set_x(self, new_x):
        self.x = new_x

    def set_y(self, new_y):
        self.y = new_y

    def get_point(self):
        return self.x, self.y


p = Point(5, 10)

print()

print(p.get_point())
```

```
p.set_x(100)

p.set_y(200)

print(p.get_point())

...

(5, 10)

(100, 200)
```

Exception handling

There are at least two distinguishable types of errors:
syntax and run time. Exceptions occur when the program
is being ran and you can handle them by using
`try/except` statements. Here's a demo of a simple
`try/except` statement in python:

```
def divide(num, den):
    try:
        x = num / den
        print('{} / {} = {}'.format(num, den, num /
den))
    except ZeroDivisionError:
        print("can't divide by zero.")


>>> divide(10, 5)

>>> divide(0, 10)
```

```
>>> divide(10, 0)

10 / 5 = 2.0

0 / 10 = 0.0

can't divide by zero.
```

The statement that wants to be executed is located within the try block. If an error occurs within the try block then the except block is executed. ZeroDivisionError is one of the many builtin exceptions in python3.

Below is another example of a try/except statement in python:

```
def import_test():
    try:
        import math
        import operating
        import sys
        print(math.pi)
        print(sys.version_info)
    except ImportError:
        print("Couldn't import something")


>>> import_test()

...

Couldn't import something
```

The reason for this is because operating is not a built-in module in python and therefore an error was triggered while in the try block. You can also use the **raise** statement to force an exception to occur. Below is an example of this in action:

```
try:
    a = input('Enter an integer ')
    raise Exception("Something strange happened")
except ValueError:
    print("An exception happened.")
```

```
>>> Enter an integer 10

Traceback (most recent call last):

File "<stdin>", line 3, in <module>

Exception: Something strange happened
```

Below is an example of a try/except/finally statement:

```
def divide(a, b):
    try:
        result = a / b
    except ZeroDivisionError:
        print("Can't divide by 0")
    else:
        print(result)
    finally:
```

```
    print('This is in the finally statement')
```

```
>>> divide(10, 2)
```

```
...
```

```
5.0
```

```
This is in the finally statement
```

The `finally` clause is executed before leaving the `try` statement. It's always executed no exceptions (no pun).

Chapter I Notes

The next three pages are for jotting down any notes you took for chapter one.

Chapter I Notes

Chapter I Notes

Chapter I Notes

Chapter II: Crafting Small Scripts, Converters, and Practical Tools in Python

In this chapter we'll learn how to build some simple yet cool python scripts that can be ran through the command line. We will learn how to make scripts that asks the user a series of questions and then process the input to do various tasks such as converting the temperature to different units, computing auto loans, and handy translators.

Project: Your Biography

Let's write a program that creates a biography for us. One way to do this is to ask a series of probing questions. Some questions that you may want to answer are things like:

- first name

- last name

- nationally

- birth place

- age

- height:

    o feet

    o inches

- weight (in pounds)

- favorite food

- favorite city

We won't use this for a dating app, we just want to use this exercise as a means to explore some of the possibilities of python ;). You can add on any additional questions you want.

Script Hints

In order to create the script follow these steps:

1) In PyCharm go to the directory where you'll place all of your programs. Right-click on the directory and select: `New → Python File`

Enter the name of the python file as *bio.py*.

2) In the PyCharm editor add a function named questions. Inside the function is where all of the statements for your logic to go inside. You can use the `input` function to read in text form the terminal. If you need to read in an `int` or a `float` then you can pass the input function into the `int` or `float` functions. For example, the following will read in a float from the terminal:

```
weight = float(input('Enter weight in lbs: '))
```

To view a list of the built-in functions in python check out this url:
`https://docs.python.org/3/library/functions.html`

3) Include the following code snippet after the `questions` function:

```
if __name__ == '__main__':
    # this is where your program starts
    questions()
```

This lets the python interpreter know where to start at. In every python file there's a __name__ variable that's set equal to __main__. Therefore, if your file explicitly includes this then it will tell the python interpreter to start here. Below is a template to how your python file will look:

```python
def questions()

    """This is the part of the

    program that prompts the user"""
if __name__ == '__main__':

  questions()
```

One of the tricky things that you may want to look out for is how to read in multiple user input in a single statement. For example, reading in a single **int** or **string** is easy because you can do something like this:

```python
>>> temperature = int(input('Temp today:'))

...

Temp today:75

>>> color = input('The color:')

...

The color:blue
```

However, what if you want the user to enter in two inputs so that you can store the data in feet and inches? One way to do that is to use the builtin string method called `split`. This will split the text around a certain character like a comma.

Solution

Below is a sample solution. Your script may have more or less questions it really depends on how you want to create it.

```
def questions():

    """This is the part of the program that prompts the user
    for a bunch of questions."""

    first_name = input('Enter your first name:
    ').capitalize()

    last_name = input('Enter your last name: ').capitalize()

    nationality = input('Enter your nationality:
    ').capitalize()

    age = int(input('Enter your age: '))

    height = input('Enter feet and inches separated by
    commas: ')

    user_input = height.split(',')

    heights = user_input[0], user_input[1]

    weight = float(input('Enter weight in lbs: '))

            favorite_food = input('Enter your favorite
            food: ').capitalize()
```

```python
        favorite_city = input('Enter in your favorite
    city: ').capitalize()

        print()

        print('First name: {}'.format(first_name))

        print('Last name: {}'.format(last_name))

        print('Nationality: {}'.format(nationality))

        print('Age: {}'.format(age))

        print('Height: {} ft {}
    in'.format(user_input[0], user_input[1]))

        print('Weight: {}'.format(weight))

        print('Favorite food:
    {}'.format(favorite_food))

        print('Favorite city:
    {}'.format(favorite_city))


if __name__ == '__main__':

    # this is where your program starts

    questions()

...

Sample input:

Enter your first name: danny

Enter your last name: hill

Enter your nationality: american

Enter your age: 47
```

```
Enter feet and inches separated by commas: 5, 5

Enter weight in lbs: 200

Enter your favorite food: pizza

Enter in your favorite city: philadelphia

Sample output:

First name: Danny

Last name: Hill

Nationality: American

Age: 47

Height: 5 ft 5 in

Weight: 200.0

Favorite food: Pizza

Favorite city: Philadelphia
```

You can run the file by opening up the terminal or command prompt and typing in the following:

```
$ python bio.py
```

Project: Temperature Converter

The world is not singular. There are many different types to something and this can vary from country to even region. Let's focus our attention on converters; there are many of these online and Google has a massive amount that you can trigger by just entering specific queries into the search box: `http://bit.ly/36e2baF`

The advantage that you have as a python programmer is that you can also add new features, create updates, and write a new script for something you can't find online. That's the beauty about knowing how to code, *you now have a new world of possibilities*.

There's three commonly used scales for measuring temperature: Celsius (°C), Fahrenheit (°F), and Kelvin (K). Celsius is used by all the countries except United States, Bahamas, Belize, Cayman Islands, and Liberia. Americans that travel abroad will probably have to go through a phase in which they have to adjust to reading temperature in Celsius… I know I did!

Kelvin is the unit of measurement used in the International System of Units (SI). The Kelvin scale is also heavily used in science and technology. Let's create a script that can convert all of the temperature units to each other and back. In order to do this we must need to know the mathematical formulas. Luckily, they're simple:

- F -> C = (F - 32) x 5/9

- F -> K = (F - 32)/1.8 + 273.15

- C -> F = (C x 9/5) + 32

- C -> K = C + 273.15

- K -> F = (K - 273.15) x 9/5 + 32

- K -> C = K - 273.15

With the above knowledge you can proceed to write your temperature conversion script. Sometimes the first step is the most difficult one to take. Here are some tips to get started.

Script Hints

We can create the function names and just include the `pass` statement so that the code doesn't do anything yet. Here's the name of the functions:

`fahrenheit_to_celsius`: Converts the user input from Fahrenheit to Celsius.

`fahrenheit_to_kelvin`: Converts the user input from Fahrenheit to Kelvin.

`celsius_to_fahrenheit`: Converts the user input from Celsius to Fahrenheit.

`celsius_to_kelvin`: Converts the user input from Celsius to Kelvin.

`kelvin_to_fahrenheit`: Converts the user input from Kelvin to Fahrenheit.

`kelvin_to_celsius`: Converts the user input from Kelvin to Celsius.

`fahrenheit_to_celsius`: Converts the user input from Fahrenheit to Celsius.

Here's a template of how the script looks:

```python
def fahrenheit_to_celsius():
    pass
def fahrenheit_to_kelvin():
    pass
def celsius_to_fahrenheit():
    pass
def celsius_to_kelvin():
    pass
def kelvin_to_fahrenheit():
    pass
def kelvin_to_celsius():
    pass

if __name__ == '__main__':
    pass
```

Solution

Below is one solution:

```python
def fahrenheit_to_celsius():
    temp_in_fahren = float(input('Enter the temperature in
Fahrenheit '))
    celsius = (temp_in_fahren - 32) * 5 / 9
    celsius = round(celsius, 4)
    print(celsius, '°C')


def fahrenheit_to_kelvin():
    temp_in_fahren = float(input('Enter the temperature in
Kelvin '))
    kelvin = (temp_in_fahren - 32) / 1.8 + 273.15
    print(kelvin, 'K')


def celsius_to_fahrenheit():
    temp_in_celsius = float(input('Enter the temperature in
Celsius '))
    celsius_to_fahren = (temp_in_celsius * 9 / 5) + 32
    print(celsius_to_fahren, '°F')


def celsius_to_kelvin():
    temp_in_cel = float(input('Enter the temperature in
Celsius '))
    celsius_to_kel = (temp_in_cel + 273.15)
    print(celsius_to_kel, 'K')


def kelvin_to_fahrenheit():
    temp_in_kelvin = float(input('Enter the temperature in
Kelvin '))
    kelvin_to_fahren = (temp_in_kelvin - 273.15) * 9 / 5 +
32
    kelvin_to_fahren = round(kelvin_to_fahren, 3)
    print(kelvin_to_fahren, '°F')


def kelvin_to_celsius():
    temp_in_kelvin = float(input('Enter the temperature in
Kelvin '))
    kelvin_to_cel = temp_in_kelvin - 273.15
    kelvin_to_cel = round(kelvin_to_cel, 3)
    print(kelvin_to_cel, '°C')
```

```python
if __name__ == '__main__':
    message = input("""Select one of the following
options:
Type 'fc' to convert from Fahrenheit to Celsius.
Type 'fk' to convert from Fahrenheit to Kelvin.
Type 'cf' to convert from Celsius to
 Fahrenheit.
Type 'ck' to convert from Celsius to Kelvin.
Type 'kf' to convert from Kelvin to Fahrenheit.
Type 'kc' to convert from Kelvin to Celsius.
Enter input here:
    """)

    # casefold is for case-insensitive comparisons

    message = message.casefold()
    if message == 'fc':
        fahrenheit_to_celsius()
    elif message == 'fk':
        fahrenheit_to_kelvin()
    elif message == 'cf':
        celsius_to_fahrenheit()
    elif message == 'ck':
        celsius_to_kelvin()
    elif message == 'kf':
        kelvin_to_fahrenheit()
    elif message == 'kc':
        kelvin_to_celsius()
    else:
        print('Not a valid option pal!')
```

You can download the script from GitHub:
http://bit.ly/36lThba

When the program runs it should display a message to show the user how to use the script. We've created a command line script so therefore some instructions on how to use it is a good start.

To do this simply create a message in the form of a string and display it. We could of used print statements, but I've opted to use a string that's wrapped in triple quotes for this as this style of strings will make it so that we don't have to worry about escaping characters like apostrophes. I'm referring to this segment of the code:

```
if __name__ == '__main__':
    message = input("""Select one of the following
options:

Type 'fc' to convert from Fahrenheit to Celsius.
Type 'fk' to convert from Fahrenheit to Kelvin.
Type 'cf' to convert from Celsius to Fahrenheit.
Type 'ck' to convert from Celsius to Kelvin.
Type 'kf' to convert from Kelvin to Fahrenheit.
Type 'kc' to convert from Kelvin to Celsius.

Enter input here:
    """)
```

When writing command line scripts we always want to make it easy for the user to enter in text. For i.e, if the user entered FC then this will not match the condition `if message == 'fc'` and therefore the program will evaluate to `False`. However, with the assistance of the `casefold` method all of these cases are evaluated so it doesn't matter what case the user enters.

Project: Auto loan Calculator

Cars are incredible mechanical inventions. Something they can be driven for countless of miles and still function properly is quite amazing. Like many quality things in life it costs money so in this project we're going to code a useful script that will help us make better purchasing decisions when deciding on a new car. The script will have two parts: one, how many months it will take to pay off the loan and two, what's the total interest paid accumulated over that period of time. Here are some formulas to get started:

$$A = P \times (1 + r) \wedge N / (1 + r) \wedge N - 1$$

Formula for calculating the accrued interest:

$$A = P(1 + rt)$$

The first formula looks tricky but it's simple once you know what all of the variables represent. Here's a quick overview:

- P = Principal or the amount owned on a loan.

- A = Total accrued amount (principal + interest).

- r = Rate of interest per year as a decimal, or interest rate / 100.

- N = Number of months in the loan period.

If you were to go into an auto company's finance department then this is the same formula that they'll use to determine your monthly car payments. In the second formula, `p(1 + rt)`, *t* represents the number of months on the loan like *n*, it's just that by using *r* and *n* together (rn) that it would be more difficult to read.

Script Hints

Create a function that calculates the monthly cost. You can break down the formula in order to minimize the chances of making a mistake. This reduces syntax errors and makes debugging potential arithmetic errors more straightforward.

Solution

Below is one solution to the problem:

```python
def monthly_cost():
    print('Gotta couple of questions for you...')
    p = float(input('Enter loan amount '))
    r = float(input('Enter interest rate (%)'))
    n = int(input('Enter loan period (in months)'))
    # convert r to a decimal and divide by interest per
year
    r = (r / 100) / 12
    # breaks formula down into 3 parts to reduce error
    # numerator
    top = r * (1 + r)**n
    # denominator
    bottom = ((1 + r)**n) - 1
    # putting it all together
    a = round(p * (top / bottom))
    # use simple interest formula
    # I = Prt
    # In this case, I = Prn
    total_interest = round(p * r * n, 3)
    print(f'Monthly costs = ${a}. Total interest =
${total_interest} ')

if __name__ == '__main__':
    monthly_cost()
```

Here's a quick breakdown of what's happening:

1) The user input is requested and stored in the variables of *p*, *n*, and *r*.

2) Once the user input is collected, the mathematics is done on separate statements to minimize errors. For example, the value of r is calculated on one line, and instead of trying to translate the mathematical formula to python code on a single line, the numerator and denominator is calculated in separate statements and then combined in this statement:

```
a = round(p * (top / bottom))
```

If you have lots of experience in python then doing everything in a single statement may be trivial. However, if you're new to the language then the important thing is to get the script to work as you can refactor (restructure) the code later.

The auto loan calculator script on GitHub:
http://bit.ly/2WzZrA0

Project: Mortgage Calculator

Getting a house is the American dream, but to obtain a dream does cost. In this exercise we're going to create a script that calculates the monthly payment that someone owes on a house. A *mortgage* is the type of loan that one takes out for a house.

Mortgage calculators can get pretty complex, so in this exercise we're going to *stupify* the process so that we just calculate the monthly payments and total mortgage that one would owe contingent on some variables. These variables are:

- **Mortgage period**: How long the mortgage will last in years.

- **Principal**: The amount of money owed on the loan.

- **Interest rate**: The percentage of principal charged by the lender for the use of their money.

Script Hints

After digging around here's the formula that you can use to calculate the monthly payments on a house:

$$p \times r \,(1 + r) \wedge N / (1 + r) \wedge N - 1$$

Put together a python script that prompts the user for their name, principal, interest rate, and then outputs the monthly payment and the total interest a homeowner would pay.

Solution

```
def mortgage():
    name = input('Enter your name ').capitalize()
    print(f"Time to calculate your mortgage payments
{name}... ")
    principal = float(input('Enter in your principal '))
```

```python
    interest_rate = float(input('Enter interest rate '))
    r = (interest_rate / 100) / 12
    n = int(input('Enter mortgage period (years) '))
    # get total number of months
    n = n * 12
    numerator = r * (1 + r) ** n
    deno = (1 + r) ** n - 1
    monthly_payment = principal * (numerator / deno)
    monthly_payment = round(monthly_payment)
    total_mortgage = monthly_payment * 30 * 12
    print(f'Monthly payment: ${monthly_payment}, total
mortgage = ${total_mortgage}')

if __name__ == '__main__':
    mortgage()
```

This script is very similar to the auto loan script except that
the formula varies:

- The variables of principal, interest rate, r, and n are
  prompted from the user.

- The numerator and denominator of the function is
  calculated separately and then combined together at
  the end to emulate the formula: p x r $(1 + r) \wedge N / (1
  + r) \wedge N - 1$

- Once the total monthly payment is calculated, the
  total mortgage is computed by taking the monthly
  payment and multiplying it by 30 and 12. The
  reason for this is because there's roughly 30 days in
  a month, and 12 months in a year.

Project: Spanish translator

Language translation is a complicated art. Even online
translation tools created by teams of talented engineers at

behemoth tech companies miss structural and linguistic elements in their translations. Therefore, to serve as a learning project we're going to simplify the process and focus on a narrow domain of words and phrases to transliterate.

Not only is this a good exercise to gain more familiarity with data structures, strings, and conditional statements, it's also a good exercise for gaining more comfort with loops in python. So, to simplify the process create a python script that focuses on translating foods and general phrases.

Script Hints

Here's an outline of the script:

```python
def food():

    pass


def general_phrases():

    pass


if __name__ == '__main__':

   print('Bienvenidos! What phases would you like to translate?')

   print('1: Common foods ')
```

```python
print('2: General phases')

your_choice = int(input("Enter your choice: '1' or '2'"))

    if your_choice == 1:

        food()

    elif your_choice == 2:

        general_phrases()

    else:

        print('Not a possible choice!')
```

The list of food items and general phrases that you want to translate is up to you. I would suggest using a dictionary to store the *Spanish-to-English* and vice-versa mappings, as this data structure is perfectly suited for this type of task.

Another thing that you may be pondering is how to create the script so that a user could easily use it? The strategy I took was that once the program runs it asks the user a question and then executes the appropriate function contingent on their feedback. Also, you may want to think about a way to normalize the text. For example, assume that the user wants to translate B*arbacoa* into English. What happens if they entered a variant spelling such as *BarBacoa* or *barbacoa*? We need a way to modify the input so that any variant of the spelling is processed. We can accomplish this by using a builtin string method such as capitalize so that the user input is in the same case as the text stored in the dictionaries.

Solution

The script contains two functions which are food and
`general_phrases`. The food function contains popular
Mexican food items written in Spanish, and if the user
types that item into the shell then its respective English
term will be printed. The `general_phrases` function
contains a list of common conversation phrases which are
translated into English when it's typed into the shell. Let's
first breakdown the food function as shown below:

```python
def food():

    """Names for some popular mexican food items.

    Translate spanish food names to english.

    """

    print('\n')

    spanish_to_english = {

        'Birria': 'Spicy stew made with goat or mutton. ',

        'Quesadilla con carne': 'season steak strips. ',

        'Barbacoa': 'Slow cooked meat in soup. Beef, goat,
or sheep.',

        'Burrito Banado': 'Wet Burrito.',

        'Huevos rancheros': '"Rancher\'s eggs." Corn
tortillas, fried eggs, topped with warm salsa.',
```

```python
    'Coctel de camarones': 'Shrimp cocktail served cold
with tomato, onion, cucumber, and cilantro. ',

    'Huevos a la mexicana': 'Eggs, tomato, onion, and
serrano chile. A classic.',

    'Huevo con Chorizo': 'Eggs and chorizo sausage.',

    'Burritos de Desayuno ': 'Breakfast burrito.',

    'Chilli con carne': 'Chili with meat.',

    'Lengua': 'Beef tongue, typically in tacos.',

    'Tripas': 'Small intestines of farm animals that\'s
cleaned, boiled, and grilled. ',

    'Al pastor': 'Pork based taco based on shawarma',

    'Suadero': 'Tender slow cooked beef brisket.
Typically served in tacos.',

    'Cabeza': 'Beef head/cheek meat. Served in soups or
tacos.',

    'Sesos': 'Brains from either a goat or cow. Popular
taco filling.'

}

print('Spanish phases available for translation:')

for spanish, english in spanish_to_english.items():

    print(spanish)

print()
```

```
translate = input('Type in spanish phase you\'ll like to
translate: ').capitalize()

for english, spanish in spanish_to_english.items():

if translate == english:

    print(spanish_to_english.get(translate))

    break

else:

    print('Word is not available for translation')
```

This code can be broken down into a couple of parts. The **spanish_to_english** dictionary contains a mapping of Mexican dishes and their explanation in English. The **for** loop iterates over the dictionary and then prints the Spanish words.

The user input is requested for the Mexican dish that they would like to translate into English. A **for** loop is used that iterates over the dictionary and then a conditional check is utilized to see if the user input matches any of the items in the dictionary. If it is then the English explanation is printed, and if not then a default message is printed letting the user know that the word is not available for translation.

Identical logic is used for **general_phrases** with the exception that the dictionary contains mappings of English to Spanish.

See the complete source code of the Spanish translator script: http://bit.ly/34fLYQf

Chapter II Recap

In this chapter we learned how to start building some real-world practical programs. Converters and calculators are some basic programs that you can start writing that allows you to start solving real world problems. If you've built a couple of these scripts then this is something that you can show your friends and family to showcase your evolving python programming prowess!

Chapter II Notes

The next three pages are for jotting down any notes you took for Chapter Two.

Chapter II Notes

Chapter II Notes

Chapter II Notes

Chapter III: How to Unlock the Power of Randomization to Create Intriguing Scripts in Python

Randomization is a phenomena that effects all humans regardless of what walk of life they're from. Many things are randomly assigned to us such as such as our date of birth, nationally, family, height, and eye color. With its inherit nonpartisan attributes, randomization is heavily used in statistics, clinical trials, and shuffling cards. Randomization is a fascinating somewhat overlooked portion of life, so let's get more familiar with it by writing some cool python programs.

Project: A Game of Dice. Humans vs. Randomization

An object that involves heavy randomization is a dice. This cubic device has a wide array of application such as tabletop games, board games, and of course gambling. A dice can easily be modeled computationally through the use of a *random* module. Python has a built in one for this which is conveniently named *random*. We'll create a simple yet elegant game that involved randomization called *A Game of Dice*.

The user will start with a bank account of $1000 and can keep making wagers on which number the dice will roll on. If the dice falls on the number that the user guesses, then the user gets the money that they wagered, and vice-versa. The game play will continue until the user runs out of money or exits from the game. This is a truly simple game that displays the beauty and the volatile nature of randomness.

Script Hints

Let's breakdown the logic:

- Who goes first? The first question is how to determine which player goes first, the human or the computer? One way is to simulate a coin flip of heads or tails. Both players are allowed to pick which side of the coin they want. If both players both pick the correct side then this process should be repeated until there's a single winner.
- How much money will each player start with? What happens when some edge cases arise like a player betting more money than they have, or enters an invalid input like a string?
- What's the minimum and maximum amount of rounds allowed?
- What happens when the player has no more money to bet?

Solution

There are a couple of things to think about when you're coding the script. For one, we need to generate a seed of numbers within the range of 1-6 that emulates the dice. We then need to figure out how many rounds we want the player to play. Here's some code to get us started:

```
from random import randint

def game():
    print('Welcome! You\'re Playing:')
    human_bank = 1000
    while True:
        you = float(input("How much money do you want
to wager? "
                          "Your bank has ${}
".format(human_bank)))
```

The first line simply imports the `randint` function from the `random` module. This allows us to generate the seed of numbers to mimic the effect of a dice. Next, a game function is created which holds the bulk of the logic. I've decided that I want to indiscriminately decide when to terminate the game-play. Therefore, a simple way to do this is to wrap our logic within an indefinite `while` loop and then include several conditional statements to control the progression of the program. Don't worry about the `while` loop not having an exiting condition as there's several ways in which we can address this later. The following code snippet shows the logic that's needed to check if the player has enough money in the bank, simulates the rolling of the dice, and then asks the user for their guess:

```
if you > 0 <= human_bank:
    print("You're betting ${}".format(you))
    dice_roll = randint(1, 6)
    human_guess = float(input('What number you
think the dice will roll on? '

                                        'Select
numbers: 1-6 '))
```

Let's analyze this portion:

```
if you > 0 <= human_bank:
```

This condition ensures that the user didn't enter in a negative number (that makes no sense) and also that the maximum amount of money that they're wagering is all the money they got in their account; anything above that shouldn't be permissible as you can't deduct more money than what is in your bank account!

Next, a random number within the range of 1-6 is generated, and then the user is asked for the amount that they're wagering. The input is *cast* or converted to a **float** which is just a decimal point number. Most people don't bet $100.50, but I've included it just for those quirky individuals that do. The next step is to update the cash in the bank account contingent on the outcome as shown in the following code snippet:

```
if human_guess == dice_roll:
    human_bank += you
    print('Dice landed on
{}'.format(dice_roll))
    print("Congrats! You now have
${}".format(human_bank))
else:
    human_bank -= you
    print("Bummers! Dice landed on
{}! Money gone...".format(dice_roll))
    print("You have
${}".format(human_bank))
if human_bank == 0:
    print("Game over! You're out of
cash")
break
```

The above code snippet includes an **if/else** statement to check to see if the user guess right or hit a big fat 0. The last **if** statement checks to see if the player's account is 0 and if it is then game over and breaks of the script.

We also need to put in some logic to determine if the player wants to continue the game or exit. The following code snippet does this:

```
                play_again = input("Enter 'y' to play
again or 'n' to stop" ).lower()
                if play_again == 'y':
                    continue
                elif play_again == 'n':
                    print("Game Over...")
                    if human_bank > 1000:
                        print("You're lucky! You won
${} ".format(human_bank - 1000))
                        break
                    elif human_bank < 1000:
                        print("Better luck next
time! You loss ${} ".format(1000 - human_bank))
                        break
                    else:
                        print("You didn't win nor
you didn't lose!")
                        break
            else:
                print('Enter a valid amount. You have
${} to bet'.format(human_bank))
game()
```

Note, if the user enters *n* for no then the game terminates and tells them if they won or lost money along with the exact amount. If the user enters *y* then the program will go on which is done through the continue statement which forces to the next cycle of the iteration.

The *Game of Dice* script is available on GitHub here: http://bit.ly/2NrRDMn

Get creative you can extend this script if you're motivated. You can make modifications such as emulating this game through a computer. In other words you can write a program in which the computer automatically wagers an amount, guess where the dice will roll on, and then simulate the dice rolling. The sequences will be quite odd but it will still be fun to watch the computer go through all of that madness.

Project: Random Person Generator

Have you ever seen those online name generators? I have, I've used them a couple of times for novels I was writing. Yes, I wrote novels under a pseudonym and occasionally had a difficult time creating names for fictitious characters! It may seem random (no pun) for a software engineer to do that, but there's actually quite a bit of parallels between writing code and writing stories. Anyhoo, to prevent from deviating let's take the functionality of a simple random name generator a step or two further. Let's write a script that allows us to randomly generate first names, last names, full names, emails, ages, telephone numbers, and email passwords. It will be a fun project to code and showcase!

Script Hints

Let's breakdown the script into individual functions so that it's more modular. Here are the functions that you can fill-in:

- `first_name`: A function that allows us to generate male or female first names. By default we'll let the program decide on the gendered pronoun.

- `last_name`: A function that allows us to generate surnames.
- `full_name`: A function that allows us to generate first and last names. Like the `first_name` function it can be gendered or we can let the program decide this.
- `age`: Generate a random age for the person. We can specify something like 1-100.
- `phone_number`: Generate a random 10 digit phone number. For the sake of simplicity we can decide on the region which in this example is North American phone numbers. The key here is that the first digit can't be 0 or 1.
- `email_password`: We can generate a random email address which uses the random person's first and last names.

There are some subtle decisions that we need to make. For one, the randomly generated full name is tied to the email that's created. Since we're creating separate functions for the `first_name, last_name`, and `full_name` functions, how can we create the email so that it's tied to the random name that's generated? There's a couple of ways to do this, for one we can call the `first_name` and `last_name` functions within the function that creates the email.

Or, we can do the second approach which is the choice I decided to do which is to include a nested function within the outer function of `full_name`. Therefore, we can create the full name and within the function give the user the option of creating an email address that uses that randomly generated full name.

If they opt for no then the full name will just be generated, and if they opt for yes then a dictionary with randomly generated name and email will be created. By the way, there's something else to think about. Where will we get the names from? You don't have to download a massive file of names, instead I just did a couple of searches online looking for popular male and female names in the USA. You're more than welcome to research popular names in any country you want. You can simply store the names in a list, and then you can use the choice function from the random module to randomly select a name as shown in the code snippet below:

```
>>> from random import choice
>>> female_names = ['Molly', 'Sue', 'Angela']
>>> choice(female_names)
...
'Angela'
```

To generate a random age is easy, you just need to generate a random number within a realistic interval:

```
>>> from random import randint
>>> [randint(1, 100) for x in range(10)]
...
[2, 23, 99, 19, 96, 27, 61, 88, 53, 38]
```

The above is simply a list comprehension that generates a list of 10 random numbers within the range of 1-100. To generate a random password here's a hint, you should investigate the functions in the string module. You have some useful things available to you in this module which are shown in the following code snippets:

```
>>> from random import choice
>>> from string import ascii_letters
>>> from string import digits
>>> from string import punctuation
>>> ascii_letters
...
'abcdefghijklmnopqrstuvwxyzABCDEFGHIJKLMNOPQRS
TUVWXYZ'

>>> digits
...
'0123456789'

>>> punctuation
...
'!"#$%&\'()*+,-./:;<=>?@[\\]^_`{|}~'

>>> [ascii for ascii in ascii_letters]
...
['a', 'b', 'c', 'd', 'e', 'f', 'g', 'h', 'i',
'j', 'k', 'l', 'm', 'n', 'o', 'p', 'q', 'r',
's', 't', 'u', 'v', 'w', 'x', 'y', 'z', 'A',
'B', 'C', 'D', 'E', 'F', 'G', 'H', 'I', 'J',
'K', 'L', 'M', 'N', 'O', 'P', 'Q', 'R', 'S',
'T', 'U', 'V', 'W', 'X', 'Y', 'Z']

>>> for x in range(5):
...     print(choice(letters))
...
H
u
K
u
A
```

Solution

Below is what I cooked up:

```python
from random import randint
from random import choice
from string import ascii_letters

def first_name(male=None, female=None):

    genders = ['m', 'f']
    male_first_names = ['Liam', 'Noah', 'William',
'Logan', 'Benjamin', 'Mason', 'Elijah', 'Oliver',
'Jacob', 'James']
    female_first_names = ['Emma', 'Olivia', 'Ava',
'Isabella', 'Sophia', 'Charlotte', 'Mia', 'Amelia',
'Harper', 'Evelyn']
    if male:
        name = choice(male_first_names)
        return name
    elif female:
        name = choice(female_first_names)
        return name
    elif male is None and female is None:
        pick_gender = choice(genders)
        if pick_gender == 'm':
            alias = choice(male_first_names)
            return alias
        elif pick_gender == 'f':
            alias = choice(female_first_names)
            return alias
```

The three functions from the import statement were previously discussed, and the `first_name` function contains the logic for randomly generating a first name.

I decided to include a list named **genders** which holds the characters *m* for male and *f* for female. Contingent on if the user entered a keyword argument, the function would randomly select a male or female first name. If no keyword argument was entered then the program would randomly select a gendered first name by passing the `genders` list into the `choice` function. Let's observe the function declaration:

```
male=None, female=None
```

The keyword arguments are both set to None. Therefore, we can check if the user passed in a value for the keyword argument if it's not equal to None. For example, let's just look at the following snippet:

```
if male:
    name = choice(male_first_names)
    return
```

The variable male will be True if it's not empty or None. If that's the case then a random male name will be generated and vice versa. Moving along, let's implement the code for generating the last name:

```
def last_name():

    last_names = ['Smith', 'Johnson', 'Williams',
'Brown', 'Jones', 'Miller', 'Martinez', 'Sanchez',
'Nguyen', 'Barnes']
    surname = choice(last_names)
    return surname
```

This is quite simple. I've just researched some popular surnames in the US on Wikipedia, added it to a list, and then implemented a random selection by using the `choice` function. Next, let's cook up the logic for creating a full name. Now, bear in mind that the route I took was to include an inner function called `create_name`. I did this because I saw where more or less the same logic was being used in two separate parts of the function. Therefore, one idea is to wrap it inside of a function and then call it when needed. The inner function I created for this purpose is called `create_name`; below is the logic for the `full_name` function:

```python
def full_name(male=None, female=None):

    email_services = ['gmail', 'aol', 'yahoo', 'aol',
'yandex', 'mail']
    create_email = input("Do you want to create an
email address? Enter 'y' for yes"
                         "or 'n' for no. ").lower()
    def create_name():
        if male:
            male_name = first_name(male='m')
            last = last_name()
            full = male_name + ' ' + last
            return full
        elif female:
            female_name = first_name(female='f')
            last = last_name()
            full = female_name + ' ' + last
            return full
        else:
            first = first_name()
            last = last_name()
            full = first + ' ' + last
            return full
    if create_email == 'y':
        email_id = {
            'name': None,
            'email': None
        }
        moniker = create_name()
```

```
        first, last = moniker.split(' ')
        e_address = (first + '.' + last + '@' +
choice(email_services) + '.com').lower()
        email_id['name'] = moniker
        email_id['email'] = e_address
        return email_id
    elif create_email == 'n':
        name = create_name()
        return name
```

Inside of the **create_name** function is similar logic that you saw in the **first_name** function. The reason for this is that it too gives the user the option of specifying a gender first-name which will be combined with the last-name. In this function the user is asked if they want to create an email or not. If they entered *y* then a dictionary will be created and returned that contains the name and email mapping pair; if not, then just the full name will be returned.

The email service like gmail, yahoo, yandex, etc, is also generated by adding it to a list, randomly selecting an item from it with **choice**, and then affixing it to the end of the email string. The following code snippet shows how to randomly generate the age of a person which can be done in one line as shown below:

```
def age():

    person_age = randint(1, 100)
    return person_age
```

That's simple stuff, just randomly generate a sequence of numbers within the range of 1-100 and then return it. Let's analyze the logic that's needed to randomly create a North American phone number:

```python
def phone_number():
    number = ''
    for x in range(1, 11):
        if x == 1:
            num = randint(2, 9)
            number += str(num)
        elif x == 4:
            num = randint(2, 9)
            number += str(num)
        else:
            num = randint(0, 9)
            number += str(num)
        if x % 3 == 0 and x <= 6:
            number += '-'
    return number
```

The first thing that we need to think of is what exactly is a valid North American phone number? It's a number that can be placed in the following format:

XNN-XNN-NNNN

In the above syntax *X* represents a number within the range of 2-9, while *N* represents a number within the range of 0-9.

Once we understand that we can create a loop that iterates over a range of 10 (the total digits in a phone number) and then randomly generate a number within the range of 0-9 EXCEPT when it's the first or fourth digit. The reason for this is because the first and fourth digits can't be within the range of 0-9, but instead within the range of 2-9; this can be accomplished with conditional checks. Also, there are two more things to keep note of. One, we need to convert the integers to strings because we want to do concatenation not sum integers. Here's an example on summing integers vs. concatenation:

```
>>> x = 5
>>> y = 10
>>> x + y
...
15
>>> str(x) + str(y)
...
'510'
```

Two, we want to format the string using hyphens to separate them. This appears simple at first glance because the string has a pattern in which the dash appears every 3 digits. If we implement a potential solution using just modulus arithmetic then the problem with this is that we could end up with a phone number that looks like this:

948-456-857-7

The reason for this is that the dash is added after every group of 3s. However, we can bypass this by adding a conditional check that says: *only add a hyphen if we are not past the sixth digit.*

This ensures that the 3rd hyphen is NOT added.

Last but not least, we must create a simple password of an arbitrary length. We can accomplish this in a single line using a list comprehension:

```python
def email_password(length=7):

    return ''.join([choice(ascii_letters) for x in
range(length)])
```

We now have several functions coded so the next step is to test them. I've created a simple function for this:

```python
def run():
    for x in range(3):
        print(first_name())
    print()
    for x in range(3):
        print(last_name())
    print()
    for x in range(3):
        print(full_name())
    print()
    for x in range(3):
        print(age())
    print()
    for x in range(3):
        print(phone_number())
    print()
    for x in range(3):
        print(email_password())
```

Here's some test output when the run function is ran through the shell assuming that the file is named random_person.py:

```
$ python random_person.py

Logan

Ava

Mia
```

Jones

Smith

Nguyen

Do you want to create an email address? Enter 'y' for yes or 'n' for no. y

{'name': 'Charlotte Jones', 'email': 'charlotte.jones@gmail.com'}

Do you want to create an email address? Enter 'y' for yes or 'n' for no. y

{'name': 'Mia Smith', 'email': 'mia.smith@yahoo.com'}

Do you want to create an email address? Enter 'y' for yes or 'n' for no. y

{'name': 'Ava Miller', 'email': 'ava.miller@yahoo.com'}

87

58

62

774-201-6881

908-307-6702

772-268-8114

MgxpMRp

kCtcLUp

VrWMVgq

You can view the full source code for *Random Person Generator:* http://bit.ly/2Pz1x1E

Project: The California State Lottery

Most states in the US allow denizens the ability to play the lottery. While I'm not advocating for folks to become chronic gamblers, I am advocating simulating the winning numbers through a python program. In this project we're going to simulate the various lotteries that California makes available which are:

- Daily 3: Pick any 3 numbers within the range of 0-9.
- Daily 4: Pick any 4 numbers within the range of 0-9.
- Fantasy 5: Pick any 5 numbers within the range of 1-39.
- Super Lotto PLUS: Pick any five numbers within the range of 1-47, and then one mega number from 1-27.
- Mega Millions: Select six numbers from two separate pool of numbers, five different numbers from 1-70, and one number from 1-25.

- Powerball: Select 5 numbers between 1-69, and one Powerball number between 1-26.

There's many ways to play the lottery. For simplicity's sake we'll give the user just one option to bet which is a straight bet, or a bet in which the player must select all of the winning numbers in the exact order in which they occur.

Script Hints

An easy way to solve this is to create a function that represents each type of lottery, and then fill in the respective logic. Once you know how to code The Daily 3, then you'll also know how to code the rest of them. The reason for this is because the logic is in essence the same, the only thing that changes are the amount of numbers to predict along with their ranges. Below are the functions to be coded:

- `daily_3`
- `daily_4`
- `fantasy_5`
- `super_lotto_plus`
- `mega_millions`
- `powerball`

Solution

Let's put the above hint into motion and code the daily_3. Here's the start:

```python
from random import randint
from time import sleep
def daily_3():
    """
    emulates the daily 3
    lottery
    """
    print("Welcome! You're playing Daily 3.")
    lucky_one = randint(0, 9)
    lucky_two = randint(0, 9)
    lucky_three = randint(0, 9)
    guess_one = int(input('Guess your lucky
number: 0 - 9 '))
    guess_two = int(input('Guess your lucky
number: 0 - 9 '))
    guess_three = int(input('Guess your lucky
number: 0 - 9 '))
```

We need to import the randint function to generate a range of random numbers, and then we could use the sleep function from the time module to pause the program for more theatricality.

The program starts off with a cordial welcome message and then three separate random variables are generated.

Next, three consecutive prompts are generated which asks the user for their guess. Once done the program can then display the results of the randomly generated numbers:

```
print("Here's the Daily 3 results...")
sleep(3)
print(lucky_one)
sleep(2)
print(lucky_two)
sleep(2)
print(lucky_three)
print([lucky_one,lucky_two, lucky_three])
```

The first random variable is printed followed by a two-second pause, and then the other two variables follow the same steps; all of the random variables are then printed as a list.

Last, the user input needs to be checked to see if it matches the randomly generated numbers. This can be easily accomplished with an **if** statement that checks each guess against the variable that holds the respective random number. If any of these evaluates to **False** then we know that the user did not win the lottery. Below is the logic for this:

```
if guess_one == lucky_one and guess_two ==
lucky_two and guess_three == lucky_three:
        print("Congrats! You won $500")
    else:
        print("Today wasn't your day! Try again.")
```

That's all there is to it! Now, fill out the remaining functions and once done you can check them against the solution here: http://bit.ly/2JCMAHR

Chapter III Notes:

The next three pages are for jotting down any notes about Chapter III.

Chapter III Notes

Chapter III Notes

Chapter III Notes

Chapter IV: Crafting Catchy Computer Art with Python Turtle

Programming is mainly an abstract endeavor, but with the help of python's turtle module you can actually see what's happening on the screen in pixel-by-pixel action! Turtle is based off the Logo programming language which was created back in 1967 at MIT as a way to make computer science accessible to kids.

The goal of this chapter is to teach you python programming concepts using this module. We'll start off by creating boring shapes such as squares and then using them to create intriguing computer artwork. I've also combined the ubiquitous concept of randomization with an open source module to allow us to randomly color artwork every time the program is ran. At the end of this chapter we'll not only have a portfolio of interesting computer artwork that we can show our friends and family, but we'll also create a fun racing game!

Geometry 101: Learning How to Create Shapes

Let's create the following image:

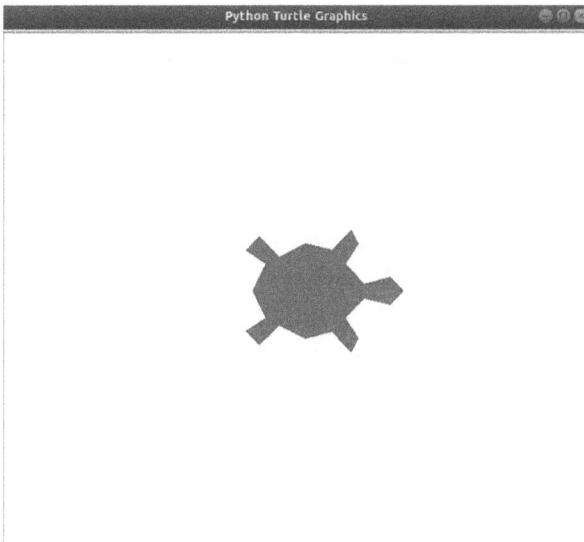

Image: Big Green Turtle

Below is the code for it:

```
import turtle
leonardo = turtle.Turtle()
leonardo.shape('turtle')
leonardo.shapesize(7.5, 7.5)
leonardo.color('green')
turtle.done()
```

The first thing we need to do is import the turtle module so that we can use all of the builtin functionality. Since it's a core library we don't need to install anything.

Once the module is imported we can go ahead and call the Turtle constructor. We can change the default shape of the turtle object to something more appealing like a green turtle using shape. Next, we call the shapesize method to manipulate its size, and then the color function to change the color to green. The last tidbit we must do is call done to start the event loop which in essence calls Tkinter's main loop function. We'll learn about Tkinter in **Chapter V** but just remember into the meantime that you need it at the end of your turtle programs.

Where to Explore All of the Cool Functionality of the Turtle Module

Before we move forward I want to mention where to go to discover the gnarly functionality of the Turtle module. The details are documented here: http://bit.ly/2BZKNbQ

Now that we know how to create a bare minimum turtle graphic along with access to its features, let's go ahead and jump to creating simple shapes.

Coding a Simple Square in Turtle

Let's reminiscence about the time our geometry teacher taught us about important things in life such as the x-y coordinate plane, how to plot, and the foundations of shapes. Remember any of that stuff? If so then great because it will serve you well when learning about the basics of turtle. Let's go back down memory lane, do you recall the essential features of a square?

Facts like that they have four sides of equal length and four right angles. We can easily create a square in turtle but we need to know a couple of things. For one, we need to know how to move forward, and we also need to know how to create angles as the interior of a square should be 90 degrees. In order to do this we need to use the builtin methods of forward and turn. Below is a demo of how to create a square in python:

```
import turtle
square = turtle.Turtle()
square.hideturtle()
square.forward(100)
square.right(90)
square.forward(100)
square.right(90)
square.forward(100)
```

```
square.right(90)
square.forward(100)
turtle.done()
```

Below is the image that it will render:

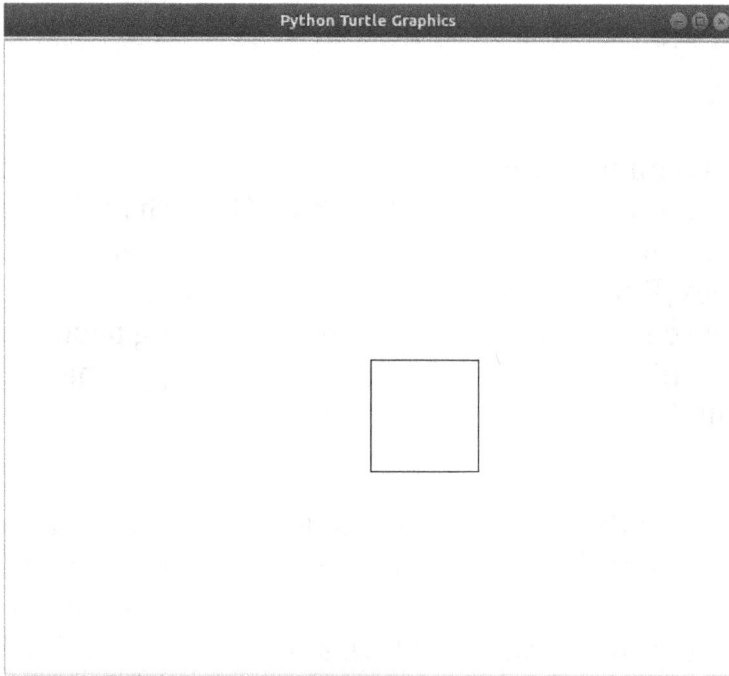

Image: Simple Square

The `hideturtle` method removes the turtle so that it won't appear in the final output, the `forward` method pushes the turtle forward an arbitrary number of pixels, and the right method pivots the turtle an arbitrary number of degrees.

There's one slight problem which is we have some duplicate code! The code pretty much has four forward and three right method calls. We can therefore add these two statements in a `for` loop as shown in the following code snippet:

```python
import turtle
square = turtle.Turtle()
square.hideturtle()
for x in range(4):
    square.forward(100)
    square.right(90)
turtle.done()
```

It literally produces the same image as above but with fewer lines of code.

But wait, we can further enhance our code by giving the user the ability to modify the square size, color, border, etc. We can wrap the logic inside of a function so that we can reuse it when needed. The below function still creates a square, albeit with the enhanced functionality:

```python
import turtle
def create_square(size=100, outer_color='black',
inner_color='blue', border=2):
    square = turtle.Turtle()
    square.pensize(border)
    square.color(outer_color, inner_color)
    square.hideturtle()
    square.begin_fill()
    for x in range(4):
        square.forward(size)
        square.right(90)
    square.end_fill()
    turtle.done()

create_square()
```

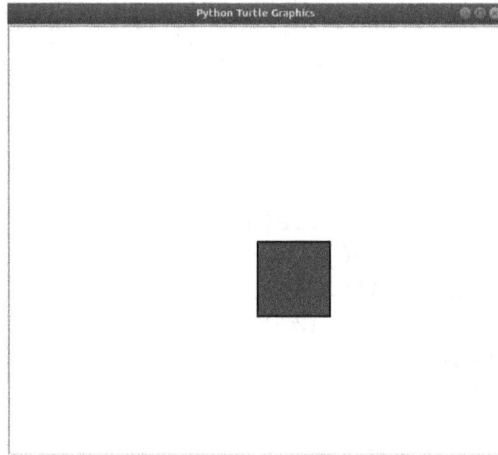

Image: Enhanced Square

The **Enhanced Square** has the same logic as the **Simple Square** with the exception that it specifies some default features of the square such as the size, outer color, inner color, and border.

If the above wasn't clear it's best to tinker with the function parameters to learn how they manipulate the square. There's a world of possibilities! This is just a sample of what's possible when you're coding with python turtle.

The PyRandomColor Open source Module

Just like how film would be humdrum with traditional black-and-white television, the same concept applies to your turtle computer graphics. Therefore, we can spice things up by adding color to our images. Turtle enables pythonistas the ability to do this via the color function.

You can read about it in the turtle module here: `http://bit.ly/2q46yo2`

You can call it with no arguments, one argument, or two arguments. When you pass in one argument it returns the `pencolor` and current `fillcolor` as a pair; it doesn't actually manipulate the color, just returns what color is in the turtle object. When you specify one argument it fills in the turtle object with that color, and when you specify two arguments the first argument is the outer color and the second one is the fill *aka* inner color.

This is great, but sometimes when you're creating images you're not exactly sure what colors to use… at least I don't! Or better yet, maybe you just want to experiment with many different types of colors until you find the combinations you like the most.

Ta-da, this is where the PyRandom*Color* module comes to the rescue! I've told you in **Chapter III** that I'm obsessed with randomization :-).

This simple module allows you to randomly pick colors which can come in handy when you want to rapidly experiment with computer art. You can access it here:
https://github.com/purcellconsult/PyRandomColor

Since the module is available via the PyPI package manager you can simply install it using pip as shown in the following code snippet:

```
$ pip install pyrandomcolor
```

Alternatively, you can clone the file to your computer using the following git command:

```
$ git clone
https://github.com/purcellconsult/PyRandomColor.git
```

It's a really simple script but it will save you the grunt work of having to think of a way to do this yourself. Here's a quick overview of the functions:

`get_random_color()`: Picks any random color.

`whites_and_pastels()`: Returns a random color that's within the white and pastel color scheme.

`grays()`: Returns a random color that's within the gray color scheme.

`blues()`: Returns a random color within the blue color scheme.

`greens()`: Returns a random color within the green color scheme.

`yellows()`: Returns a random color within the yellow color scheme.

`browns()`: Returns a random color within the brown color scheme.

`oranges()`: Returns a random color within the orange color scheme.

`pinks_violets()`: Returns a random color within the pinkish and *violetish* color scheme.

Here's an example on how to use the module to randomly color a square:

```
from random_colors import get_random_color
import turtle
def random_color_square1():
    square = turtle.Turtle()
    square.hideturtle()
    square.pensize(3)
    square.pencolor(get_random_color())
    square.color(get_random_color(),
get_random_color())
    square.begin_fill()
```

```
for x in range(4):
    square.forward(100)
    square.right(90)
square.end_fill()
turtle.done()


random_color_square1()
```

Below is the output image:

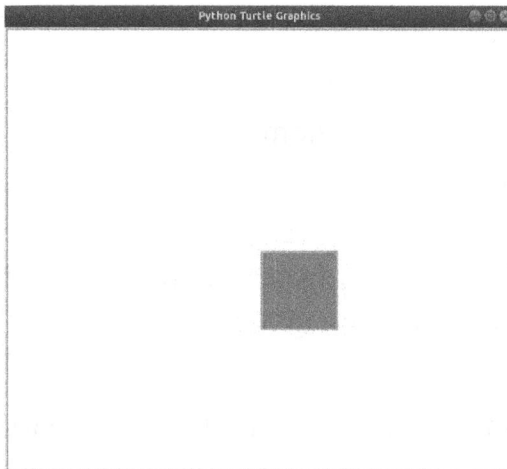

Random Color Square One

Note, since this uses randomized colors there's a high probability that there will be some variance compared to the image you generate. This is why I think this so cool, every execution of the script will most likely generate something slightly different.

Here's a quick explanation. To randomly fill in the inner and outer colors of a turtle object you need to do two things: use the color function and then pass in the get_random_color function twice from the random_colors module. Remember, when you specify two arguments in the color function, the first one is the outer color while the second one represents the inner color. To get the color to generate we need to include the begin_fill method call before the loop and the end_fill method after it.

Here's another example of what you can do with *PyRandomColor*.

```
def random_color_square2(cycles=10):
    square = turtle.Turtle()
    go, turn = 100, 90
    square.hideturtle()
    square.pensize(3)
    square.pencolor(get_random_color())
    square.color(get_random_color(), get_random_color())
    for x in range(cycles):
        for y in range(4):
            square.forward(go)
            square.right(turn)
        go -= 10
    turtle.done()
random_color_square2()
```

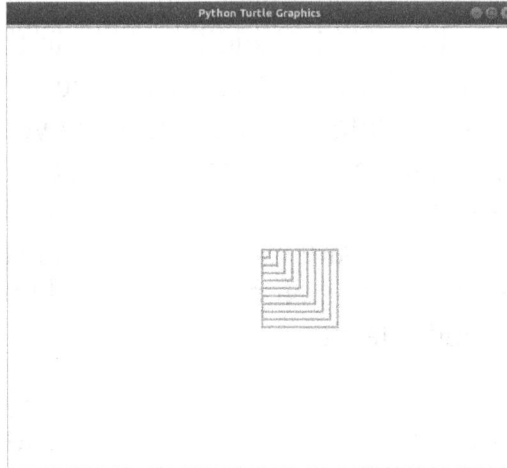

Image: Random Color Square Two

How to Craft Beautiful Designs from Ugly Shapes with Turtle and PyRandomColor

Putting together simple shapes is an easy way to draw complex and surprisingly stunning patterns. Let's start with what we already know which is the square. We can use that simple shape to create a multi-colored cube as shown in the following code snippet:

```
import turtle
from random_colors import get_random_color
square = turtle.Turtle()
```

```
square.hideturtle()
for x in range(4):
    square.color(get_random_color(),
get_random_color())
    square.begin_fill()
    square.right(90)
    for y in range(4):
        square.forward(100)
        square.right(90)
    square.end_fill()
turtle.done()
```

Below is one way in which the image could look:

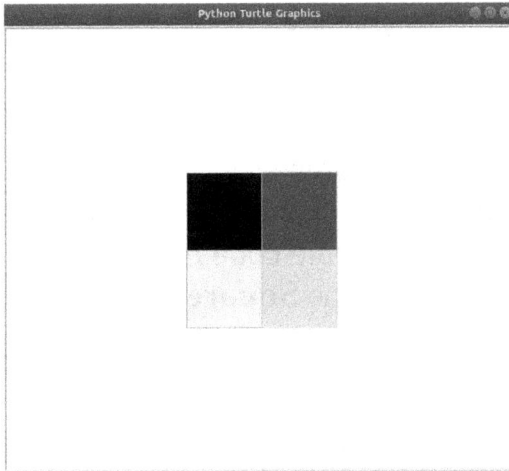

Image: Square Cube

A nested loop is used to create four squares of varying colors. After each iteration the new square is rotated 90 degrees and then the new square is created. Can you guess what happens if the inner loop iterated two cycles instead of 4? Here's how it would manipulate the image:

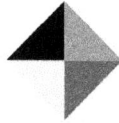

Image: Triangle Cube

It will be a cube of multicolored triangles instead of squares. We can continue to create more complex images by building on top of the simple code for creating a square. Let's make some tweaks to the **Square Cube** code as shown below:

```
import turtle
from random_colors import get_random_color
square = turtle.Turtle()
square.hideturtle()
square.speed(0)
for x in range(50):
    square.color(get_random_color(),
get_random_color())
    square.begin_fill()
    square.right(35)
    for y in range(4):
        square.forward(100)
```

```
        square.right(90)
    square.end_fill()
turtle.done()
```

If we do that then we can get an interesting image as
shown below:

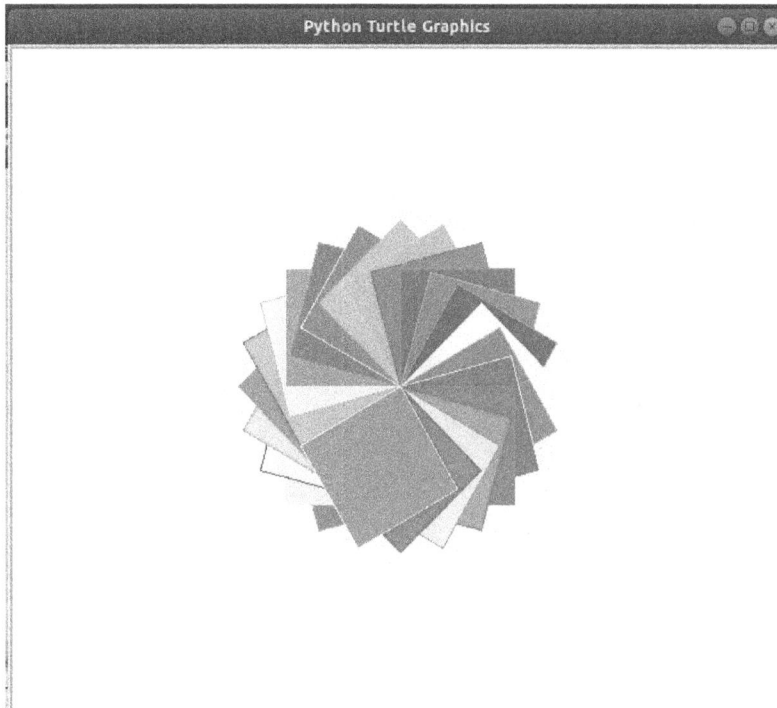

Image: Square Artwork

This image looks complex but the code is straightforward.
It's an image that contains 50 squares which are rotated by
35 degrees after each complete inner loop cycle. Go ahead
and play with the code to see how a couple of tweaks to

these simple turtle algorithms can create derivative computer artwork.

A Randomly Generated Night Sky Sprinkled With Stars

The sky dotted with stars is a beautiful sight to see. The good news is that you don't have to be in the Grand Canyon to stargaze as you can bring the sky to your computer screen as shown in the image below:

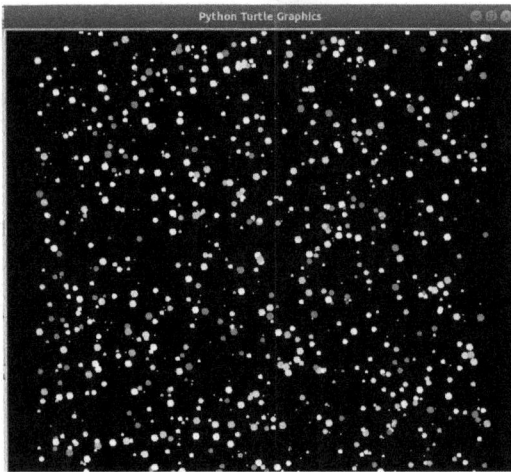

Image: Night Sky

To emulate the effect of the star dotted sky we need two things: a black background and a whole bunch of white-and-grayish circles of various sizes spread throughout. To re-create this image we need to first learn how to make circles which fortunately is simple with turtle. Here's the method signature for this:

```
turtle.circle(radius, extent=None, steps=None)
```

The radius is the distance from the center of a circle to any point on its circumference. The *extent* if not given draws the entire circle, and if given draws an arc. The arc is drawn in a counter clockwise direction if the radius is positive, otherwise it's drawn in a clockwise direction. Let's learn how to create a simple circle and then a simple arc. The following code snippet shows a full circle:

```
import turtle
x = turtle.Turtle()
x.hideturtle()
x.pensize(2)
x.circle(100)
turtle.done()
```

Below is an example of how the circle looks in python:

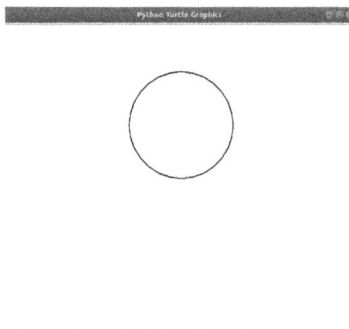

Image: Circle

As you have probably figured out the bigger you make the radius the bigger the circle, and vice-versa. To simulate an arc pass in a keyword argument in the circle method as shown below:

```python
import turtle
x = turtle.Turtle()
x.hideturtle()
x.pensize(2)
x.circle(100, extent=150)
turtle.done()
```

Below is the generated image:

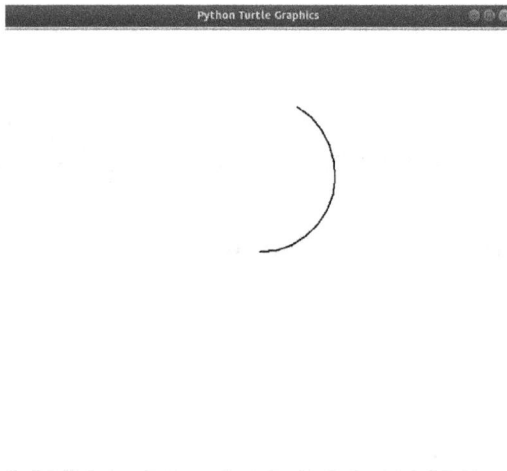

Image: Arc

Now, with our newfound knowledge of circles we have one more thing that we need to know. We need to know how to move turtle objects around the screen. The reason for this is that we don't want to plot all of the circles at the exact position because doing so will lead to a blob of circles around the same area which looks kind of awkward. Instead, we want to randomly draw the circles at various locations on the screen in order to get a better distribution. We can accomplish this by using three more methods from the turtle class:

- penup: Picks the pen up, no drawing happens.
- pendown: Puts the pen down.
- goto(x, y=None): Move the turtle to the x and y coordinates.

The penup and pendown methods operate in a similar fashion to how you draw on a piece of paper. Imagine that you're chilling at your desk drawing a masterpiece on a blank piece of paper. What would happen if you decide to pick your pen up and move it to a different position on the paper? The pen will now appear at that spot. When you put the tip of your pen back down this is when the drawing happens.

This is important to use in turtle because if you don't use these methods and instead just use the goto method then lines will be drawn along the path which is not what you want. In other words, it's like if you're sketching a stick figure on a piece of paper and instead of picking the pen up to draw the individual parts like the head, chest, and arms, you instead keep the pen down the entire time which will lead to ugly squiggly lines.

Lastly, we need to set the background color of the screen which can be done by using the `bgcolor` method from the `Screen` class. From there we can call the `bgcolor` method in order to modify the color. Let's code the night sky in turtle by first including all of the imports:

```
import turtle
from random import randint
from random_colors import whites_and_pastels
```

We'll import the turtle module in order to use its functionality, the `randint` function from random to generate random numbers which will be used for positioning, and then the `white_and_pastels` function from the `PyRandomColor` module in order to give the stars various shades of white and gray to mimic the sky. Below is the beginning of the function definition:

```
def create_night_sky(stars=1000):
    sky = turtle.Turtle()
    sky.speed(0)
    turtle.Screen().bgcolor('black')
```

Here we're just creating a function with a default argument of 1000. This simply represents how many stars will be drawn in the final image. Next, the turtle constructor is called and the speed is set to 0 so that the images will be drawn at the fastest rate possible. The `bgcolor` method from the `Screen` instance is called to set the background of the screen to black which will give us the *night sky* effect. Now, here's the rest of the function:

```
num = stars
for x in range(num):
    sky.color(whites_and_pastels())
    sky.begin_fill()
```

```
    sky.penup()
    sky.goto(randint(-300, 300), randint(-300, 300))
    sky.circle(randint(1, 5))
    sky.pendown()
    sky.end_fill()
turtle.done()
```

Here's where the stars are created and then moved each iteration. We create a `for` loop which iterates by default 1000 cycles. From there, we set the colors of the stars. To do this we again need to utilize the `color, begin_fill`, and `end_fill` methods. Since we want to arbitrarily determine the position of the stars, we can use the `penup, goto`, and `pendown` methods.

Remember, the `penup` method is used to pick the pen up so that no lines follow when the position is being allocated. We then pass the `randint` function into the x and y arguments of `goto` to randomly determine the x/y coordinate of where the circle will be displayed.

Once the position is located we can use the `circle` method to create the actual circle. Since we want to simulate stars in the sky, we need to ensure that some stars are bigger than others. How can we accomplish this? We need to arbitrarily set the radius of the circle by again using the `randint` function. The last statement in the program must be `turtle.done` so that the final image will stay; without this the generated image goes *bye-bye*. We can then run the program to see the sky populated with varying sized stars in various locations. Every time the program is ran the output will be different but it could be difficult to notice without *screen-shotting* and comparing the pictures due to the small size of the circles.

More Circle Artwork with Turtle

While we're on the topic of using circles to create cool graphics we can again use these shapes for more creative computer art. Just like we did in the previous example, let's again create random circles of various sizes. However, this time we're going to give them any of the various random colors from the `get_random_color` function in the `PyRandomColor` module. The code for crafting this is listed below:

```python
import turtle
from random_colors import get_random_color
from random import randint

def create_color_circles(amount=500):

    cir = turtle.Turtle()
    cir.speed(0)
    quantity_circles = amount
    length = cir.getscreen().window_width()
    height = cir.getscreen().window_height()
    for cycles in range(quantity_circles):
        cir.penup()
        x, y = randint(-length, length), randint(-
height, height)
        cir.goto(x, y)
        cir.pendown()
        rad = randint(5, 50)
        cir.begin_fill()
        cir.color(get_random_color())
        cir.pendown()
        cir.circle(rad)
        cir.end_fill()
    turtle.done()
```

Here's the output:

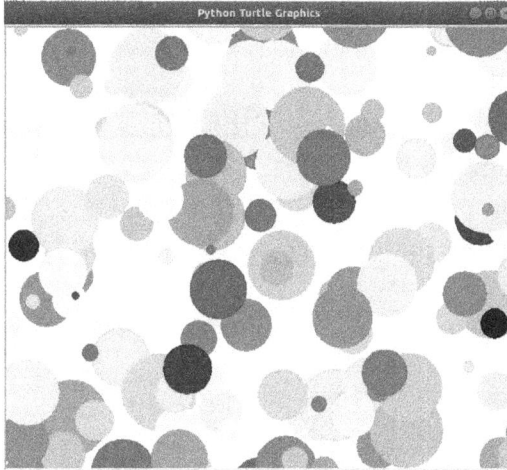

Image: Random Colored Circles

Now, let's dig through the code. This function is similar to the code used to create the **Night** image.

However, there are two simple tweaks. One, the circles are populated across the whole dimension of the user's screen. This is accomplished by using the `getscreen` method which returns the `TurtleScreen` object that the turtle is drawing on. The size of the screen is stored in the `length` variable and the height is stored in the `height` variable. A loop is created that by default iterates 500 times and then repeats the following steps:

- Picks up the pen

- Randomly sets location by selecting a variable x within the range of -length to length, and a y within the range of -height to height
- Puts the circle at the location of the x and y variables
- Places the pen at its new position with goto. No more squiggly lines!
- Randomly sets the size of the radius within the range of 5 through 50 pixels
- Calls the begin_fill method
- Calls the get_random_color method from PyRandomColor
- Puts the pen down
- Draws the circle
- Calls the end_fill method
- Calls turtle.done()

Pretty cool stuff. Let's continue riding the circle art bandwagon. Let's go ahead and create a simple *spirograph* with a circle. Below is the code snippet for this:

```python
import turtle
from random_colors import get_random_color

def create_circles(cycles=100):

    turtle.bgcolor('black')
    turtle.pensize(3)
    turtle.speed(0)
    for i in range(cycles):
        turtle.color(get_random_color())
        turtle.circle(125)
        turtle.right(25)

turtle.done()
```

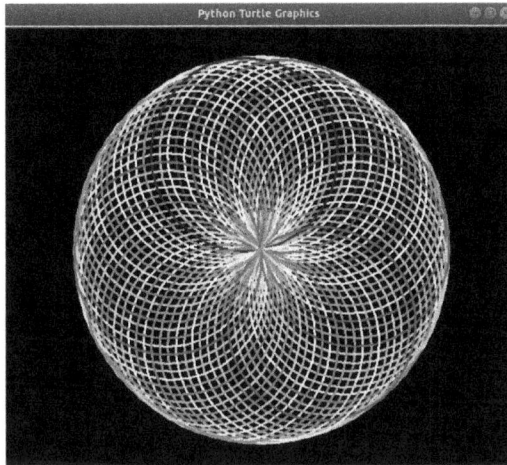

Image: Spirograph 1

So we in essence create a screen with a black background and then use a `for` loop to iterate a predetermined amount of cycles. From there, we create the random colored circle, and then rotate the pen right 25 degrees. We can play with the angle rotation to see how much it affects the output. If we want the circles to appear in a linear fashion instead of having them rotate then we can make a couple of adjustments.

One, we don't need to rotate the circle after each cycle, but instead we need to move it forward by a predetermined amount. If we want to start the drawing at the area located furthest to the left then we can get the width of the screen. Inside of the `Screen` module is a method called `window_width` that allows you to get the width of the screen.

It's important that we take the negative of that value and then store it in a variable. We want it to be negative because just like with the x-y Cartesian coordinate system the negative value denotes to the left side on the coordinate system. To get the *left most* portion of the screen we can do something like the following:

```
x = -(turtle.Screen().window_width())
```

After each iteration we can update the x value and then move the turtle pen to the new location using the penup, pendown, and goto methods. Below is the updated code snippet:

```python
import turtle
from random_colors import get_random_color
def create_circles(cycles=30):

    turtle.bgcolor('black')
    turtle.pensize(3)
    turtle.speed(0)
    x = -(turtle.Screen().window_width())
    for i in range(cycles):
        turtle.penup()
        turtle.goto(x, 0)
        turtle.pendown()
        turtle.color(get_random_color())
        turtle.circle(100)
        x += 50
    turtle.done()
```

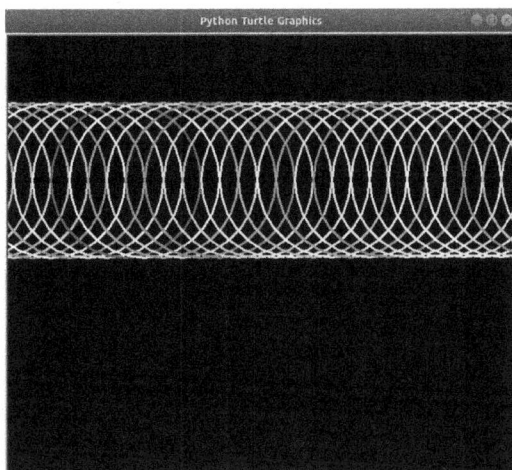

Image: Circle Art

Creating Party Lights and Delicious Fruit Candy Pieces with Simple Lines

To create a line in python all you have to do is just move forward with the turtle object. You could also optionally add width to the pen so that the line will have more *thickness*. For example, the following code consists of several lines that rotate 45 degrees each iteration around the origin and kind of mimics a snow flake as shown below:

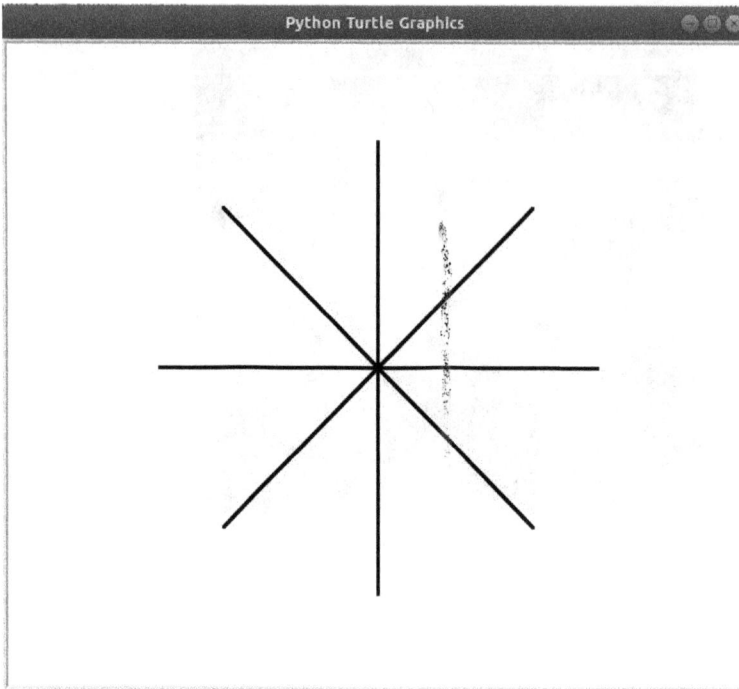

Image: Lines

Here's the code snippet for doing this:

```python
import turtle
line = turtle.Turtle()
line.hideturtle()
line.pensize(3)
for x in range(8):
    line.forward(200)
    line.goto(0, 0)
    line.left(45)
turtle.done()
```

Since we now know how to create lines the next thing to do is to create some computer artwork with them using turtle! Let's do something cool by following this simple algorithm:

- Generate many lines

- Use the `PyRandomColor` module to randomly color each line

- Add some thickness to the lines

- Randomly allocate the x/y coordinates of each line

So, pretty much more of the same stuff that we saw previously. Below is how the image looks:

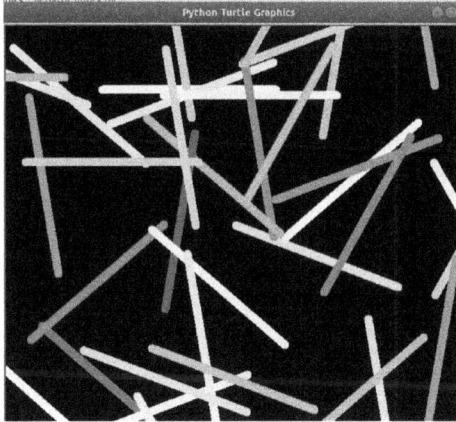

Image: Party Lights

Below is the code for this which is similar to the code used to create the **Night** and **Random Colored Circles**:

```
import turtle

from random_colors import get_random_color
```

```python
from random import randint

def create_lights(number=70, angle=20):

    width = turtle.Screen().window_width()

    height = turtle.Screen().window_height()

    screen = turtle.Screen()

    screen.bgcolor('black')

    screen.screensize(width, height)

    turn = angle

    lights = turtle.Turtle()

    lights.speed(0)

    lights.hideturtle()

    for x in range(number):

        x, y = randint(-height, height), randint(-width, width)

        lights.pensize(12)

        lights.pencolor(get_random_color())

        lights.right(turn)

        lights.forward(250)

        lights.penup()

        lights.goto(x, y)

        lights.pendown()
```

```
turtle.done()
```

There's no new logic here, the only thing you haven't seen yet is how to control the thickness of the lines. This can be done by calling the `pensize` method.

We can in essence take the above logic and manipulate the image slightly by modifying the `pensize` aka thickness, and the `forward` function, aka length of the line. Here's how the modified loop portion of **Party Lights** look:

```
for x in range(number):

    x, y = randint(-height, height), randint(-width, width)

    lights.pensize(30)

    lights.pencolor(get_random_color())

    lights.right(turn)

    lights.forward(75)

    lights.penup()

    lights.goto(x, y)

    lights.pendown()
```

Here's the generated image:

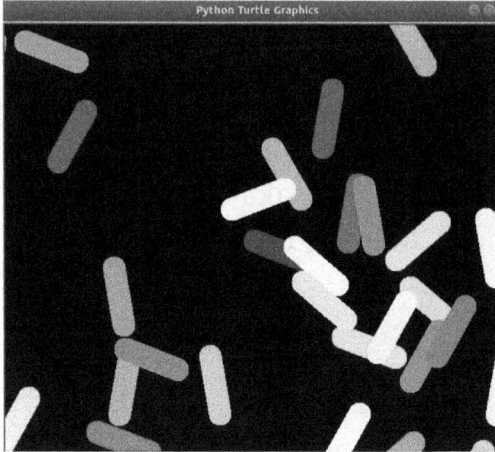

Image: Candies

Same logic, we just made a couple of tweaks to the pensize and forward methods.

Project: 4 Little Turtles Racing Game

Let's put our newfound knowledge of the turtle module to the test and build ourselves a simple little game. This game is called *4 Little Turtles* and it's actually more of a simulator than a game, but yet we can still get a ton of entertainment value from this. Below are the rules to 4 Little Turtles:

a) Create a game that consists of 4 little turtles with unique colors.

b) Include a countdown that starts the game.

c) Create a graphical finish line with the text "Finish Line" directly above it.

d) Randomly generate the speed of the turtles.

e) Write an algorithm to accurately determine which turtle crossed the finish line first.

f) Write a message to let the user know which turtle won.

g) Write a simple algorithm to make the winning turtle spin and then grow in size afterwards.

h) Have fun playing and showing the game off to your friends and family!

Below are some screenshots of how the game looks:

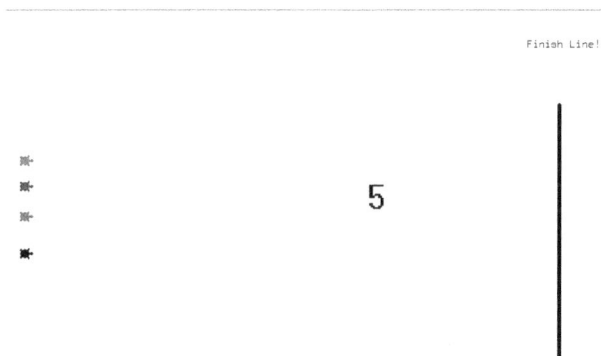

Finish Line!

5

Image: 4 Little Turtles Screenshot 1

Finish Line!

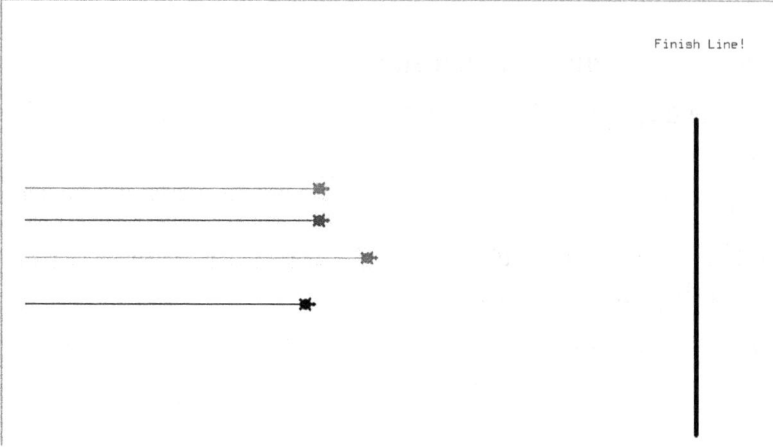

Image: 4 Little Turtles Screenshot 2

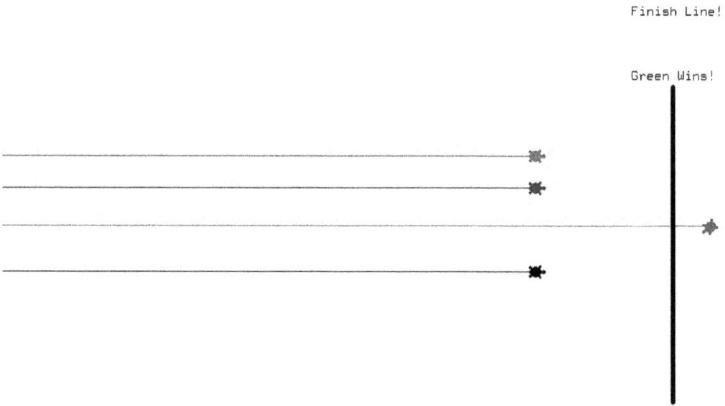

Finish Line!

Green Wins!

Image: 4 Little Turtles Screenshot 3

Code Cool Stuff With Python

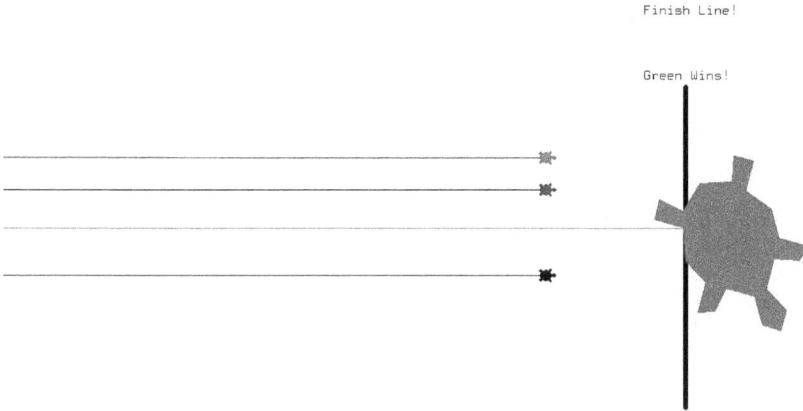

Finish Line!

Green Wins!

Image: 4 Little Turtles Screenshot 4

Script Hint

Compared to the other programs we created so far I felt that this program would fare better by using the *object oriented* approach. I felt that this style of programming would make the code base cleaner, more modular, and easier to extend when needed. Below is the list of methods that's used in the script:

__init__: The constructor which creates the four turtle objects along with getting the height and width of the screen.

`start_turtles`: Sets the default positions of the turtles.

`finish_line`: Draws the finish line in the racing game and aligns it on the right hand portion of the screen.

`countdown_timer`: Starts the countdown timer for the script starting at 5. Use the sleep function from the time module to pause the program.

`set_turtle_speeds`: Randomly sets the speed of the turtles within the range of 1-25.

`turtle_race`: Contains the logic that gets the turtles to start moving towards the finish line.

`victor_dance`: Contains the logic that gets the winning turtle to do a celebratory dance after they win and then grow in size.

`if __name__ == '__main__':` We need a way to start the game. I've decided to use this approach. It can be used for larger projects to define a starting point of execution of the project. Python doesn't have an official main method like C or Java, so you can use this to control the flow of execution of a script.

Solution

We can break down the solution by analyzing the methods. We'll start at the top and work our way to the bottom. Let's start with the class declaration and then the constructor or __init__ method:

```python
import turtle
from random import choice
from time import sleep
class TurtleRace:

    def __init__(self):

        self.red = turtle.Turtle(shape='turtle')
        self.blue = turtle.Turtle(shape='turtle')
        self.green = turtle.Turtle(shape='turtle')
        self.black = turtle.Turtle(shape='turtle')
        self.height = turtle.Screen().window_height()
        self.width = -turtle.Screen().window_width()

        self.red_x = self.width + (-self.width/10)
        self.blue_x = self.width + (-self.width/10)
        self.green_x = self.width + (-self.width/10)
        self.black_x = self.width + (-self.width/10)

        self.red_y = self.height - (self.height / 1.20)
        self.blue_y = self.height - (self.height / 1.1)
        self.green_y = self.height - (self.height / 1)
        self.black_y = self.height - (self.height / .9)

        self.red_speed = 0
        self.blue_speed = 0
        self.green_speed = 0
        self.black_speed = 0
        self.finish = turtle.Turtle()
```

The __init__ method which acts as the constructor is used to create and initialize the variables that will be used throughout the program. The variables created within the __init__ method are prefaced with the self convention and belong to the object instance. These types of variables are known as **instance variables** and are included within the methods of a class.

The variables are made to represent each color of turtle. The dimensions of the screen is captured and stored in the `height` and `width` variables. Since many computer screens come in various dimensions the actual value of `height` and `width` are contingent on the screen that the user is on. The way to get these variables in turtle is to use the `window_height` and `window_width` methods.

Next, the starting positions of the turtles are set. The x position is set equal to the width plus `-self.width/10`.

This seems like an odd computation, but without adding extra pixels to the width the turtle won't be on the screen! I've decided to add `-self.width/10` to it so that hopefully it scales gracefully across various computer screens. All of the turtles start the race at the same x coordinate.

It's important to note that the width of the screen will be a positive number like 600. However, since we want the turtle to start furthest to the left, then what we need is to make sure that the width is negative. That's done in this code snippet here:

```
self.width = -turtle.Screen().window_width()
```

The same logic is used to create the y position of the turtles, but instead we get the height of the screen. The speeds of the turtles are initially set to 0, and the finish variable is created which will control the positioning and creation of the finish line graphic.

There will be more variables created later on but the variables declared within the constructor will be manipulated by other methods; the variables created later will only be used locally.

Now that we got the positioning of the turtles right, the next step is to go ahead and set them up. Here's the code for this:

```python
def start_turtles(self):
    self.red.penup()
    self.red.color('red')
    self.red.goto(self.red_x, self.red_y)
    self.red.pendown()

    self.blue.penup()
    self.blue.color('blue')
    self.blue.goto(self.blue_x, self.blue_y)
    self.blue.pendown()

    self.green.penup()
    self.green.color('green')
    self.green.goto(self.green_x, self.green_y)
    self.green.pendown()

    self.black.penup()
    self.black.color('black')
    self.black.goto(self.black_x, self.black_y)
    self.black.pendown()
```

The turtle color is defined and then it's moved to its default location on the screen using the `goto` method. Let's move on to the portion of the code that shows how to position and draw the finish line.

```python
def finish_line(self):

    self.finish.hideturtle()
    self.finish.speed(0)
    self.finish.pensize(7)
    self.finish.hideturtle()
    self.finish.penup()
    self.finish.goto(-self.width / 2, self.height / 3)
    self.finish.pendown()
    self.finish.right(90)

    for cycles in range(30):
        self.finish.forward(cycles)

    self.finish.penup()
    self.width = int(self.width / 2.4)
    self.finish.goto(-self.width, self.height / 2)
    self.finish.pendown()
    self.finish.write('Finish Line!', font=('Verdana', 13))
```

The only somewhat tricky part is figuring on the correct x/y values to set the line at. I used the width and height of the screen as a reference point, and then scaled it by dividing by an integer. Let's investigate this portion of the code here:

```python
self.width = int(self.width / 2.4)
```

```python
self.finish.goto(-self.width, self.height / 2)
```

What's happening is that the width is being divided, and then as you can see the x portion has a negative symbol in front of it. Since the default value of the `width` is negative (was set this way in the constructor), when another negative is added in front of it then it becomes positive.

The `for` loop creates the finish line which is simply a line that's moved forward and then rotated to the right 90 degrees so that it's vertical. To increase the thickness of the line just increase the value that's passed into the `pensize` method.

In order to get the turtle object to write text above the finish line we simply use the same `x` value, but we divide the height by a smaller number so that the text will appear above the line instead of overlapping with it.

Below is the code snippet that modifies the font face:

```python
self.finish.write('Finish Line!', font=('Verdana', 13))
```

You can modify the tuple to change the font face and size.

Now, let's inspect the logic for creating the countdown script for the game:

```python
def countdown_timer(self):

    secs = 5
    while secs > 0:
        message.write(secs, font=('Verdana', 50))
        secs -= 1
        sleep(1)
        message.clear()
```

I've decided to countdown the game starting at 5, but you can of course increase this number to something bigger or smaller if desired. Since the game counts down a total of 5 seconds, the loop will cycle 5 times and during each iteration the sleep function is called; the 1, pauses the program for a second. Since the condition is while secs > 0, make sure to decrement the counter by 1 each cycle so that it will indeed reach 0. The next step is to randomly set the speeds of the turtles:

```python
def set_turtle_speeds(self):

    speeds = [gait for gait in range(1, 25)]
    self.red_speed = choice(speeds)
    self.blue_speed = choice(speeds)
    self.green_speed = choice(speeds)
    self.black_speed = choice(speeds)
```

To make this game unpredictable the turtles will be randomly assigned a speed from 1 to 25. A turtle that gets assigned a low speed like 1 will truly mimic a turtle in real life!

Let's next dig into the logic required to make the turtle race and then determine the actual winner. Below is the beginning snippet for this:

```python
def turtle_race(self):

    race_on = True
    while race_on:
        if self.red.pos()[0] >= self.finish.pos()[0]:
            if self.blue.pos()[0] >=
self.finish.pos()[0] and self.blue.pos()[0] >
self.red.pos()[0]:
                self.finish.penup()
                self.finish.goto(-self.width,
self.height / 2.9)
                self.finish.pendown()
                self.finish.write('Blue Wins!',
font=('Verdana', 13))
                race_on = False
                return self.blue
            elif self.green.pos()[0] >=
self.finish.pos()[0] and self.green.pos()[0] >
self.red.pos()[0] :
                self.finish.penup()
                self.finish.goto(-self.width,
self.height / 2.9)
                self.finish.pendown()
                self.finish.write('Green Wins!',
font=('Verdana', 13))
                race_on = False
                return self.green
            elif self.black.pos()[0] >=
self.finish.pos()[0] and self.black.pos()[0] >
self.red.pos()[0]:
                self.finish.penup()
                self.finish.goto(-self.width,
self.height / 2.9)
                self.finish.pendown()
                self.finish.write('Black Wins!',
font=('Verdana', 13))
                race_on = False
                return self.black
        self.red.forward(25)
```

```
            self.finish.penup()
            self.finish.goto(-self.width, self.height /
2.9)
            self.finish.pendown()
            self.finish.write('Red Wins!',
font=('Verdana', 13))
            race_on = False
            return self.red
```

A method is defined and a while loop is created with the flag **race_on** which is initially set to `True`. The `while` loop will keep cycling indefinitely until **race_on** is set to `False` which happens when the winning turtle crosses the finish line. The next question is how do we determine that?

Every turtle object that we've drawn up until this point has an x and y position, just like on the Cartesian coordinate system. We can access the current x or y value of a turtle object by invoking the `pos` method. So, if we want the current x value of the red turtle we can do this:

```
self.red.pos()[0]
```

If we want the y position of the red turtle we can do that:

```
self.red.pos()[1]
```

However, we need something to compare it to. Since we want to know which turtle won we should compare the current x value of each turtle to the x value of the finish line. That's where this snippet of code comes in handy:

```
if self.red.pos()[0] >= self.finish.pos()[0]:
```

At the end of each iteration we need to progress the turtles forward which can be accomplished by using the following code snippet:

```
self.red.forward(self.red_speed)
self.blue.forward(self.blue_speed)
self.green.forward(self.green_speed)
self.black.forward(self.black_speed)
```

You would think that this is all we needed to determine the winning turtle but there's something we're missing. For example, let's assume that the red turtle speed was 20, and let's also assume that the red turtle has crossed the finish line. While the red turtle may indeed cross the finish line, there are a couple of edge cases we need to check. In order to fully crown Mr. or Mrs. Red Turtle the winner, we need to check that the other three turtles don't have a speed that's greater than 20.

If they do, then what could happen is that our program has a bug in it because the red turtle could be declared the winner when it doesn't actually have the highest speed.

Let's still assume that the red turtle again has a speed of 20, and let's assume that the blue turtle has a speed of 22. The speeds are really close, and if we don't add some type of conditional check then what could happen is that the red turtle could be declared the winner simply because it's placement in code comes before that of the blue turtle.

That's where those nested conditional checks come into play. For example, let's analyze this code snippet:

```python
if self.red.pos()[0] >= self.finish.pos()[0]:
    if self.blue.pos()[0] >= self.finish.pos()[0]
and self.blue.pos()[0] > self.red.pos()[0]:
```

What this says in English: *if the x value of the red turtle is greater than the x value of the finish line; and if the position of the blue turtle is greater than the x value of the finish line and the x position of the blue turtle is greater than the x position of the red turtle then...*

The code that follows positions the pen above the finish line, sets `race_on` to `False` which makes the `while` loop stop, and then returns the blue turtle. We need to return the winning turtle as that decides the victory dance. However, we need to do these conditional checks for the green and black turtle since the red turtle appears first in the code. Only after those checks are done is it safe for us to assume that the red turtle is indeed the winning turtle.

At this point I decided to include `forward(25)` on all of the winning turtles so that it will look like they sprint past the finish line like in an actual track race.

Similar logic applies to the blue, green, and black turtles with the exception that they need to check the turtles that come before them in the code. That's the logic to start the turtles off racing and to ensure that the turtle with the fastest speed is crowned the winner.

The next code snippet shows the logic on how to make the turtle do its silly spin dance and then grow in size afterwards:

```python
def victor_dance(self):

    if self.turtle_race() == self.red:
        for x in range(50):
            self.red.right(90)
        self.red.shapesize(10, 10)
        return
    elif self.turtle_race() == self.blue:
        for x in range(50):
            self.blue.right(29)
        self.blue.shapesize(10, 10)
        return
    elif self.turtle_race() == self.green:
        for x in range(50):
            self.green.right(30)
        self.green.shapesize(10, 10)
        return
    elif self.turtle_race() == self.black:
        for x in range(15):
            self.black.right(90)
        self.black.shapesize(10, 10)
        return
```

As you can see it's simply a `for` loop that cycles a variable number of times and then rotate the turtle to the right a variable number of degrees. At the end of the iteration the `shapesize` method is invoked which allows the change in dimensions of the turtle. The first parameter coincides with the stretch width while the second parameter coincides with the stretch length.

The last portion of the code is to just execute all of the methods. The order does matter because we don't want to call a method that's trying to access an object that doesn't exist. Below is the code for this:

```
if __name__ == '__main__':
    message = turtle.Turtle()
    message.hideturtle()
    race = TurtleRace()
    race.start_turtles()
    race.finish_line()
    race.countdown_timer()
    race.set_turtle_speeds()
    race.turtle_race()
    race.victor_dance()
    turtle.done()
```

That's all there is to The *4 Little Turtles* game.

You can download the complete script from GitHub: http://bit.ly/331MBN9

What cool interesting things have you created with the turtle module? Post on social media about it with the hashtag: *#codecoolstuff*

Chapter IV Notes

The next three pages are for jotting down any notes about chapter IV.

Chapter IV Notes

Chapter IV Notes

Chapter IV Notes

Chapter V: Building Practical Desktop apps in Python Using the Core Tkinter Library

Tkinter is the de facto standard GUI in python. It's wired into its core meaning that if you already have python installed then you can simply import the library and start coding away. The Tkinter library is actually a binding to the Tk GUI toolkit.

Tk is a free and open source cross platform widget toolkit that provides a library full of an assortment of GUI widgets in various programming languages. There are other GUI tool kits for python but I've decided to focus on just one in this chapter, and figured that focusing on the one that shipped with python makes the most sense.

There's a lot of functionality in Tkinter, so instead of highlighting a bunch of boring jargon the game plan will be to build something simple first, analyze it, and then work through some more interesting projects. So, let's get started by building a simple GUI program!

A Simple Tkinter GUI

Here's a very basic Tkinter app:

```python
import tkinter as tk
root = tk.Tk()
root.title('Simple Tkinter App')
root.geometry('500x500')
tk.mainloop()
```

Here's the output:

Image: Simple Tkinter GUI

Woohoo, you've created your first Tkinter app! While this app doesn't do anything useful it does illustrate some key concepts with Tkinter.

```python
import tkinter as tk
```

This statement simply imports the `tkinter` module. We could have done import `tkinter` but we save ourselves some extra typing by shortening the text to `tk`. Also, some developers prefer to import everything from `tkinter` by doing the following:

```
from tkinter import *
```

However, since we're still learning the ins-and-outs of tkinter it's a bit premature.

```
root = tk.Tk()
```

This starts the tcl/tk interpreter under the cover. Then, the `tkinter` commands are translated into tcl/tk commands. The main window and the interpreter are linked, and both are needed in order for the `tkinter` app to work. So, in other words, creating an instance of Tk initializes the interpreter and creates the root window. Next...

```
root.title('Simple Tkinter App')
```

The title method allows you to set the title of the app. Nothing tricky!

```
root.geometry('500x500')
```

This sets the dimensions of the window AND to position the window within the user's Desktop. The default size is typically too small so to modify it to the way you like use this method.

In order to specify the dimensions it has to be in the 'widthxheight' format. If you want to specify the location of the app then pass in the arguments after the dimensions in the format x, y. You have to put a plus or minus in front of the values as indicated in the code snippet below:

```
root.geometry('500x500+10-10')
```

The plus value for the x coordinate pushes the window to the left while the minus pushes it to the right; for the y value the + value pushes the window up while the minus value pushes the window down.

Last but not least is the call of the `mainloop` method which keeps your application running. It functions similar to an infinite loop that you may use when you're creating a game script.

This code is more or less boilerplate that you can commit to memory so you at least know the basics of setting up your app. The problem with the app is it doesn't process user input and therefore can't interact with the user which negates a crucial benefit of GUIs. Therefore, let's create another silly app which has an input form which asks the user for their name and then displays it. In order to do this we need to know just 3 more things:

We need some way to provide instructions to a user, in the way an email form for example may have the text *email* next to the form, or the text *first name* next to the form in which you enter in your name. We can accomplish this using the Label widget. More on this later...

We need a way so that the user can enter in data into the program. What we need is known as an Entry widget. More on this later...

We need a way to process the user input. We can do this yet again with the help of another widget, this time the Button widget. More on this later...

Ok, before moving forward we need to understand what widgets are so let's describe it, and then get to coding the *Name App*.

The Label, Entry, and Button Widgets

Tkinter supports 15 widgets. A widget is an object that provides various controls that the coder can use to integrate with the GUI.

A label as the name suggests allows you to display text or an image on the screen. Below is an example on how to create one:

```
name = tk.Label(root, text='Your favorite
food?').pack()
```

So, we call the `Label` method using `tk`, and then pass in the root which is the parent of the class.

We can then make use of the various attributes in the label widget and text is one of them. As the name indicates it allows you to add text to the label. It's important to note that the size of the label will automatically be set so that all of the characters in the text attribute are displayed.

However, there's an attribute that allows you to use the width and height of the label. Below is a listing of some of the attributes for the label widget.

- `width`: Change the width of the label widget. If you set this too small then all of the text won't show.

- `height`: Adjust the height of the label.

- `bg`: The background color.

- `fg`: The foreground color.

- `bd/borderwidth`: The width of the label border.

- `padx`: Extra horizontal padding to add around the text.

- `pady`: Extra vertical padding to add around text.

- `textvariable`: Associates a Tkinter variable, typically a `StringVar` with the label.

A good resource that you can use for additional help is: http://effbot.org/tkinterbook/label.htm

The **pack** method is one of several geometry managers. This will yet be explained later on, but the important thing to know is that without it your label widget won't show up on the GUI. Next, let's briefly learn about the `Entry` label.

This allows you to collect data from the user in the GUI. Similar to the way in which you'll use the input function from the python core, you can also use this widget to collect data from the user. Below is an example on how to use the Entry widget in python:

```python
first_name = tk.Entry(root).pack()
```

It has all of the attributes that were previously discussed with label except padx/pady.

Now, if we want text to appear in the Entry widget itself then there's no attribute that we can place within the Entry widget. Instead, we need to create what's called a StringVar variable, and then pass it in as the text variable value of the Entry widget.

A StringVar is in essence part of the variable classes in tkinter; there's also the BooleanVar, DoubleVar, and IntVar classes. These serve as wrappers for their respective data type.

Enough discussing, let's look at some code snippets:

```python
default_value = tk.StringVar()
```

In order to use StringVar you can use tk to call it. From there, you can use some of the various functions inside of it such as set which sets the default value of this variable, or get which retrieves the value stored.

Here's an example on how to use the set method in python:

```
default_value.set('pizza?')
```

In order to make the text `'pizza'` show up in the Entry widget you must add the variable name associated with `StringVar` to the `Entry` widgets text variable as shown below:

```
first_name = tk.Entry(root,
textvariable=default_value).pack()
```

On to the Button widget. Below is an example that not only shows how to create a button in Tkinter but also shows how to tie together the pieces of using a `Label`, layout manager, `StringVar`, `get/set`, and an `Entry` widget. Here's the code for the simple greetings app:

```
import tkinter as tk
root = tk.Tk()  # starts a tcl/tk interpreter under the
cover
root.title('Greetings App')
root.geometry('350x250')
def greeting():
    greetings = tk.Label(root, bg='tan', borderwidth=3.5,
relief='groove', text="Hello {} :-)! Welcome to the
wonderful\n"
                                "world of
programming. Enjoy your stay!".format(default_value.get()))
    greetings.pack()
default_value = tk.StringVar()
default_value.set('????')
first_name_label = tk.Label(root, text='What\'s your
name?').pack()
first_name = tk.Entry(root,
textvariable=default_value).pack()
button = tk.Button(text='Click Me!',
command=greeting).pack()
tk.mainloop()
```

The following screenshots illustrates how the GUI works:

Image: Greetings app 1.

Here's the sample output:

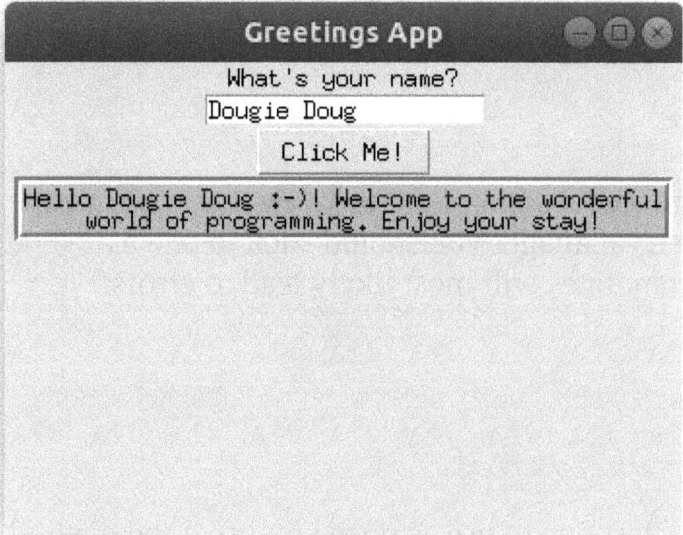

Image: Greetings app 2.

As again, not a ground shattering app but it does its job with providing insights into the possibilities of Tkinter. There are a couple of more things that we should understand until we move onto more advanced Tkinter apps.

Next, we'll learn the geometry manager in Tkinter and the various options for controlling the layout of items in your GUI. Afterwards, we'll cover the remaining widgets and build some simple yet practical apps that can provide insights into what we can accomplish with Tkinter.

The 3 Ways to Order Widgets in Tkinter

There are three ways in which you can control the layout of items in your GUI, they are:

- pack
- grid
- place

Before we proceed it's important to know that when you pick a geometry manager you should stick to one as mixing different ones will most likely lead to errors.

Pack

This is the easiest one to use as once you use it the method takes care of the ordering itself. It does this by positioning the elements relative to each other. Here's an example of the pack geometry layout manager in action:

```python
import tkinter as tk
root = tk.Tk()
root.title('Tinker Geometry Managers')
colors = ['black', 'red', 'orange', 'blue', 'green',
          'yellow', 'brown', 'gold']
# The Label geometry layout manager in Tkinter
label_one = tk.Label(text='The Black Label').pack()
label_one_black = tk.Label(root,
bg=colors[0]).pack(fill=tk.X)
label_two = tk.Label(text='The Red Label').pack()
label_two_red = tk.Label(root, bg=colors[1]).pack(fill=tk.X)
label_three = tk.Label(text='The Orange Label').pack()
label_three_orange = tk.Label(root,
bg=colors[2]).pack(fill=tk.X)
label_four = tk.Label(text='The Blue Label').pack()
label_four_blue = tk.Label(root,
bg=colors[3]).pack(fill=tk.X)
label_five = tk.Label(text='The Green Label').pack()
label_five_blue = tk.Label(root,
bg=colors[4]).pack(fill=tk.X)
label_six = tk.Label(text='The Yellow Label').pack()
label_six_yellow = tk.Label(root,
bg=colors[5]).pack(fill=tk.X)
label_seven = tk.Label(text='The Brown Label').pack()
label_six_brown = tk.Label(root,
bg=colors[6]).pack(fill=tk.X)
label_eight = tk.Label(text='The Gold Label').pack()
label_six_gold = tk.Label(root,
bg=colors[7]).pack(fill=tk.X)
root.mainloop()
```

Here's the output:

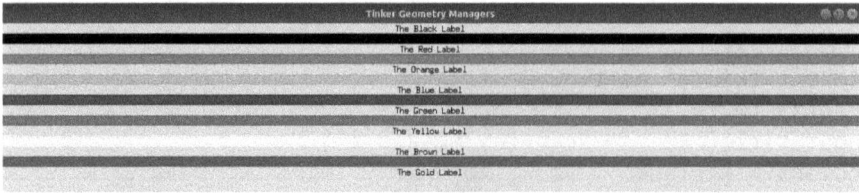

Image: Pack Geometry Manager.

As you can see from the above code snippet, the position of the labels are automatically determined. Nothing new added from this code snippet that you haven't seen before except for the background colors in the labels. Next is an example of the place geometry layout manager.

Place

The place layout manager allows you to do absolute and relative positioning with tkinter. So, you can specify exactly where you want a widget to appear in the GUI. Here's a code example that shows how to use the place geometry layout manager:

```python
import tkinter as tk
root = tk.Tk()
root.title('Tinker Place Geometry Manager')
root.geometry('375x350')
colors = ['black', 'red', 'orange', 'blue', 'green',
          'yellow', 'brown', 'gold']
width, height = 0, 0
for x in range(len(colors)):
    tk.Label(text=colors[x], width=10).place(x=0, y=height)
    tk.Label(bg=colors[x], width=15).place(x=100, y=height)
    height += 15
tk.mainloop()
```

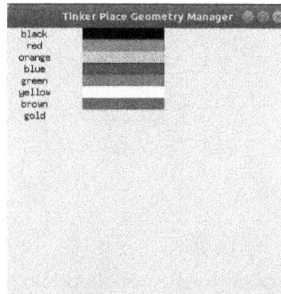

Image: Place Geometry Example.

Grid

The last layout manager to explain is the grid. It organizes components of the GUI by placing them in a 2-dimensional table which consists of rows and columns. Widgets with the same row and column number will be placed on the same line. The size of the grid doesn't have to be determined because the grid manager automatically takes care of this for you. The below code snippet utilizes the grid manager:

```python
import tkinter as tk
root = tk.Tk()
root.title('Tkinter Grid Geometry Manager')
```

```
root.geometry('875x200')
colors = ['black', 'red', 'orange', 'blue', 'green',
          'yellow', 'brown', 'gold']
i = 0
for x in colors:
    tk.Label(text=x, width=15).grid(row=i+1, column=i)
    tk.Label(bg=x, width=15).grid(row=i, column=i)
    i += 1
tk.mainloop()
```

Here's how the GUI looks:

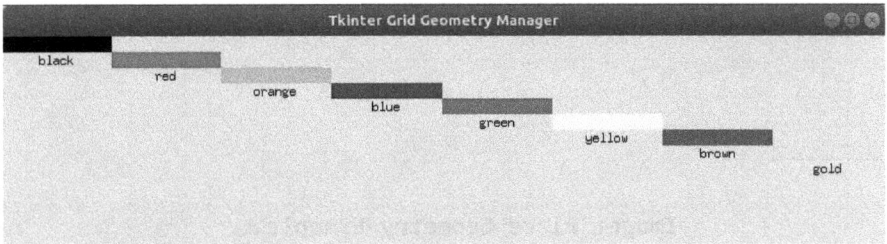

Example: grid layout manager example.

Before we move on there's something that I want to get off my chest. I think it's time that we outta convert our procedural GUI programs into their OOP counterparts.

How to Convert Procedural GUI to Their OOP Counterpart

The benefit of converting the procedural styled GUI into OOP is that in essence the program will be more modular and therefore makes debugging more precise. Let's take the very first GUI that we created in this chapter and then convert it to OOP style. To do this we need to accomplish 5 things:

1) We need to create a class.
2) We need to create a constructor using the __init__ method.
3) We need to prepend the self convention in front of the variables.
4) We need to create an instance of the class.
5) We need to call the mainloop method.

Below is the **Simple Tkinter GUI** we created earlier in this chapter converted into OOP:

```
import tkinter as tk
class HelloWorld:
    def __init__(self):
        self.root = tk.Tk()
        self.root.title('Simple Tkinter Class Example')
        self.root.geometry('500x500')
if __name__ == '__main__':
    app = HelloWorld()
    tk.mainloop()
```

From now on this is the style we will use to create our GUIs.

Now that we've got the basics of Tkinter under control the next step is to put all of our newfound knowledge to use by building some simple yet practical Tkinter apps. We'll convert a command line program that we created in **Chapter II** into a GUI program and also work through two more interesting projects.

Project: Temperature Converter GUI

Here's a screenshot of how the app looks:

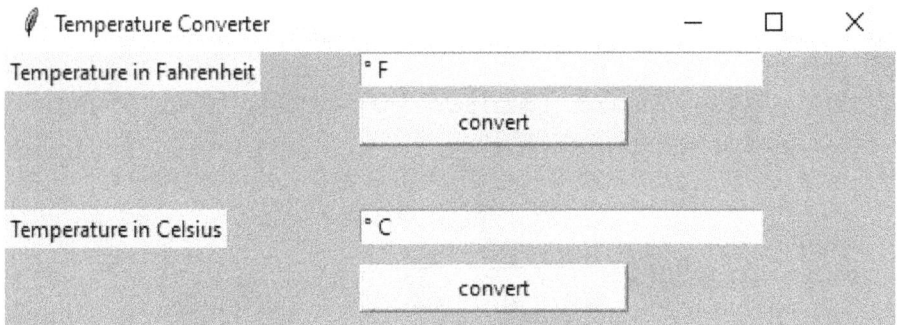

Image: Temperature Converter GUI.

Script Hint

__init__ : The constructor will initialize the instance variables and do the initial setup for the program. In here a TK instance will be created and the title, size, and the background color set. The instance variables that will be used throughout the program are created here.

fahrenheit_to_celsius: This is the method that will calculate the conversion of Fahrenheit to Celsius.

celsius_to_fahrenheit: This is the method that will convert from Celsius to Fahrenheit.

create_widgets: In this method we'll create and configure the various widgets that will be used throughout such as the Label, Button, and Entry widgets.

if __name__ == '__main__': This is the entry point of the program. An instance of the class will be created and then the various methods will be invoked all followed by a call to `mainloop`.

Now that we have the script broken down into methods, we can go ahead and start writing the program.

Solution

Here's the initial portion of the script.

```python
import tkinter as tk
from tkinter import Tk, Label, Button, Entry, StringVar
class Temperature:
    def __init__(self):
        self.root = Tk()
        self.root.title('Temperature Converter App')
        self.root.geometry('500x150')
        self.root.configure(bg='tan')
        # Gets the user input
        self.get_fahren = StringVar()
        self.get_celsius = StringVar()
        # Sets the user input. By default
        # its default values are ° F and ° C
        self.get_fahren.set('° F')
        self.get_celsius.set('° C')
```

As you can see tkinter as well as the various modules that will be used are individually imported. That way, we don't have to invoke `tk` each time we want to use them. The `get_fahren` and `get_celsius` variables will be used to get the input that the user entered into the program. The following methods will take the input that the user entered and then convert it to its appropriate temperature scale:

```python
def fahrenheit_to_celsius(self):
    """
    Fahrenheit to Celsius formula is:
    (x - 32) * 5/ 9
    """
    try:
        celsius = float(self.get_fahren.get())
        result = (celsius - 32) * 5 / 9
        result = str(round(result, 3)) + '° C'   # rounds float to 3 places, then converts to string
        self.get_fahren.set(result)
    except ValueError as e:
        self.get_fahren.set('ERROR!!!')
```

We must first get the value that the user entered into the program which is done in this line of code:

```python
celsius = float(self.get_fahren.get())
```

The `get` method allows us to get the value that the user entered, convert it to a float, and then execute the rest of the equation. Once the equation is executed the value is set. The `try/except` statement catches any non-numerical value that's trying to be added to the equation. Below is the `celsius_to_fahrenheit` equation which is more or less the same logic except with a different equation:

```python
def celsius_to_fahrenheit(self):
    """
    Celsius to Fahrenheit formula is:
    F = (x x 9/5) + 32
    """
    try:
        fahrenheit = float(self.get_celsius.get())
        result = (fahrenheit * 9 / 5) + 32
        result = str(round(result, 3)) + '° F'
        self.get_celsius.set(result)
    except ValueError as e:
        self.get_celsius.set('ERROR!!!')
```

The following method creates the widgets that will be used and then positions them using the place layout manager.

```python
def create_widgets(self):
    """
    arranges the labels,
    Entry, and buttons.
    """
    fahrenheit_to_celsius_label =
Label(text='Temperature in Fahrenheit').place(x=0, y=0)
    fahrenheit_to_celsius_widget =
Entry(text='enter temperature',
textvariable=self.get_fahren, bg='white
smoke').place(x=200, y=0, width=225)
    fahrenheit_to_celsius_button =
Button(text='convert', width=15, border=2,
command=self.fahrenheit_to_celsius)
    fahrenheit_to_celsius_button.place(x=200, y=25,
width=150)
    # celsius to fahrenheit label, widget, and
button
    celsius_to_fahrenheit_label =
Label(text='Temperature in Celsius').place(x=0, y=85)
```

```
    read_celsius_widget = Entry(text='enter
temperature', textvariable=self.get_celsius, bg='white
smoke').place(x=200, y=85, width=225)
    celsius_to_fahren_button =
Button(text='convert', width=15, border=2,
command=self.celsius_to_fahrenheit)
    celsius_to_fahren_button.place(x=200, y=115,
width=150)
```

Note, that with the place layout manager you can set the absolute positioning of the widgets. Remember, the `Label` widget just allows us to display text or images on the screen, and the `Entry` widget allows us to collect or display a single line of text. The `Entry` widgets have a keyword argument of `textvariable` which is set to `self.get_fahren` and `self.get_celsius` for the appropriate `Entry` widgets. This allows us to get the user input, and then the temperature conversion is done when the appropriate button is clicked. The button executes the appropriate method that does the conversion of Fahrenheit to Celsius and vice-versa. Remember, the buttons can execute the appropriate method when clicked on by using the `command=self.some_method` as a keyword argument. The last step is to set everything up and to then run the program. This is done in the following code snippet:

```
if __name__ == '__main__':
    app = Temperature()
    app.create_widgets()
    tk.mainloop()
```

The `if __name__ == '__main__'` conditional is started which serves as the entry point to the program, and then an instance of the class is created along with a call to the `create_app` method. Lastly, the `mainloop` method is called which emulates an infinite loop and keeps the GUI active until closed.

Project: BMI Calculator App

Body mass index or BMI for short is the acronym for **body mass index**. It's a standard for estimating a healthy weight for individuals. The formula works by taking the body mass of an individual and dividing it by the square of the body height. We'll design a GUI that accepts the height and weight of an individual in the following units:

- Height in feet
- Weight in pounds
- Height in meters
- Weight in kilograms

Below is a screenshot of the finished BMI Calculator that we'll be building:

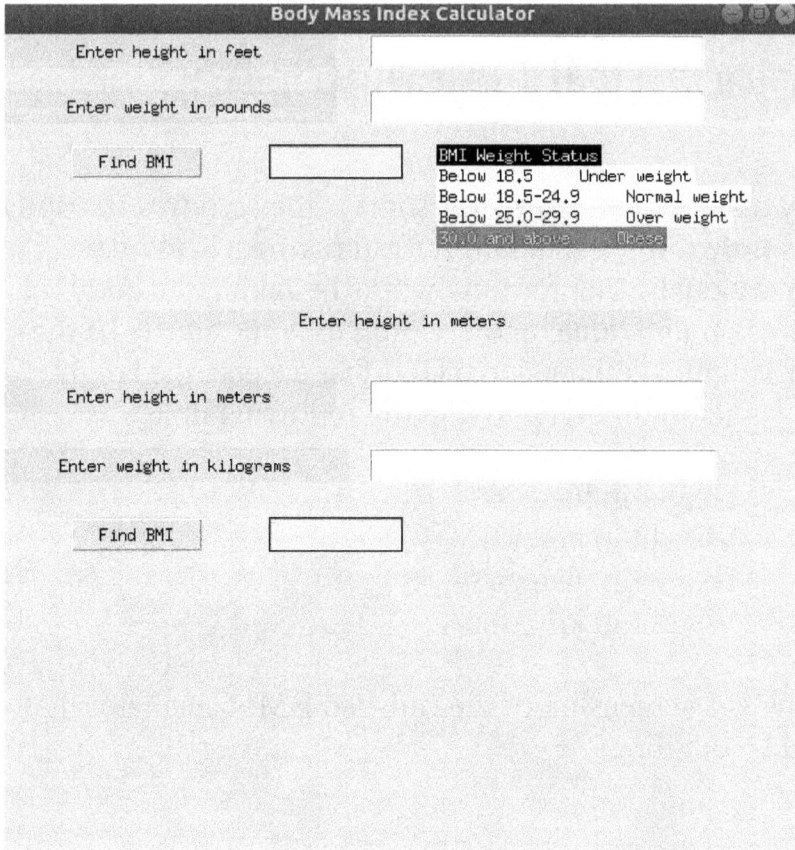

Image: Body Mass Index Calculator Screenshot.

Script Hint

We can use a similar template to create this GUI for the one we used to create the Temperature Desktop app. Here's an outline of methods that you can use to create the app:

__init__: Here you can initialize the variables that will be used throughout the program.

create_widgets: This method creates and places the Label, Entry, and Button widgets on your GUI.

weight_in_pounds: Accepts the weight in pounds and height in feet and then calculates the BMI.

weight_in_kilograms: Accepts the weight in kilograms and height in meters and then calculates the BMI.

run: This is the method that will contain all of the setup steps for the GUI.

Solution

Let's get the basic stuff out of the way such as the class and __init__ method:

```python
class BMICalculator:
    def __init__(self, master):
        self.height_in_feet = tk.StringVar()
        self.weight_in_pounds = tk.StringVar()
        self.pounds_bmi = tk.StringVar()
        self.height_in_meters = tk.StringVar()
        self.weight_in_kilos = tk.StringVar()
        self.kilos_bmi = tk.StringVar()
```

The above are just the variables that will be manipulated throughout the program. Let's next create all of the widgets that will be used throughout the program and wrap them inside a method:

```python
def create_widgets(self):
    bmi_feet_inches_label = Label(text='Enter height in
feet')
    bmi_feet_inches_label.place(x=0, y=0, height=25,
width=250)

    bmi_feet_inches_entry =
Entry(textvariable=self.height_in_feet)
    bmi_feet_inches_entry.place(x=275, y=0, height=25,
width=250)
    bmi_feet_inches_entry.config(highlightbackground='white
smoke')

    height_inches_label = Label(text='Enter weight in
pounds')
    height_inches_label.place(x=0, y=40, height=25,
width=250)

    weight_pounds_entry =
Entry(textvariable=self.weight_in_pounds)
    weight_pounds_entry.place(x=275, y=40, height=25,
width=250)
    weight_pounds_entry.config(highlightbackground='white
smoke')

    bmi_button = Button(text='Find BMI',
command=self.weight_in_pounds)
    bmi_button.place(x=50, y=80, width=100)

    display_bmi_inches = Entry(textvariable=self.pounds_bmi)
    display_bmi_inches.place(x=200, y=80, height=25,
width=100)
    display_bmi_inches.config(highlightbackground='black',
bg='lavender')

    Label(bg='black', fg='white', text='BMI Weight
Status').place(x=325, y=80)
    Label(bg='floral white', text='Below 18.5      Under
weight').place(x=325, y=95)
    Label(bg='azure', text='Below 18.5-24.9      Normal
weight').place(x=325, y=110)
```

```
    Label(bg='yellow', fg='black', text='Below 25.0-29.9
Over weight').place(x=325, y=125)
    Label(bg='red', fg='white', text='30.0 and above
Obese').place(x=325, y=140)

    meters_and_kilo_label = Label(text='Enter height in
meters')
    meters_and_kilo_label.place(x=100, y=200, width=400)
    height_meters_label = Label(text='Enter height in
meters')
    height_meters_label.place(x=0, y=250, height=25,
width=250)
    height_meters_entry =
Entry(textvariable=self.height_in_meters)
    height_meters_entry.place(x=275, y=250, height=25,
width=250)
    height_meters_entry.config(highlightbackground='white
smoke')

    bmi_meters_kilo_label = Label(text='Enter weight in
kilograms')
    bmi_meters_kilo_label.place(x=0, y=300, height=25,
width=260)
    bmi_meters_kilo_entry =
Entry(textvariable=self.weight_in_kilos)
    bmi_meters_kilo_entry.place(x=275, y=300, height=25,
width=260)
    bmi_meters_kilo_entry.config(highlightbackground='white
smoke')

    bmi_button_meters = Button(text='Find BMI',
command=self.weight_in_kilograms)
    bmi_button_meters.place(x=50, y=350, width=100)

    display_bmi_meters = Entry(textvariable=self.kilos_bmi)
    display_bmi_meters.place(x=200, y=350, height=25,
width=100)
    display_bmi_meters.config(highlightbackground='black',
bg='lavender')
```

The `Labels`, `Entry`, and `Button` widgets are created
here. Let's focus on the `config` method; it's called on the
`Entry` widgets and allows you to set the background color
of that widget. Let's inspect this portion of the GUI:

Image: BMI Weight Status Labels.

These are simply labels that correspond to this portion of the code:

```
Label(bg='black', fg='white', text='BMI Weight
Status').place(x=325, y=80)
    Label(bg='floral white', text='Below 18.5     Under
weight').place(x=325, y=95)
    Label(bg='azure', text='Below 18.5-24.9     Normal
weight').place(x=325, y=110)
    Label(bg='yellow', fg='black', text='Below 25.0-29.9
Over weight').place(x=325, y=125)
    Label(bg='red', fg='white', text='30.0 and above
Obese').place(x=325, y=140)
```

The next step is to create the methods that do the BMI computation contingent on the values:

The appropriate values entered into the GUI are retrieved by calling the `get` method on `self.height_in_feet` and `self.weight_in_pounds`. The formula is then executed and the result is stored in `pounds_bmi`. The `weight_in_kilograms` method is the same logic with the exception of a different formula being used:

```python
def weight_in_kilograms(self):

    meters = float(self.height_in_meters.get())
    kilograms = float(self.bmi_meters_kilo_entry.get())
    bmi = round(float(kilograms / (meters ** 2)), 2)
    self.kilos_bmi.set(bmi)
```

It's time to test that everything finally works. In this app we're going to take a different approach to running the file compared to the Temperature app. What we're going to do is create a function that does all of the setup for the GUI and then within the `if` conditional calls the `run` function. Below is how the `run` function looks:

```python
def run():
    root = tk.Tk()
    root.title('Body Mass Index Calculator')
    root.geometry('600x600')
    root.configure(bg='AntiqueWhite2')
    app = BMICalculator(root)
    app.create_widgets()
    tk.mainloop()
```

The run function calls the `Tk` constructor which allows us to use the various functionality built into the package such as setting the title, size, and configuring the `bg` color for the GUI. From there we create an instance of the class and pass in the root and then call the `create_widgets` method belonging to the `BMICalculator` class, and the `mainloop` method to get the GUI to constantly run. From there we can use the `if __name__ == '__main__'` functionality to create the entry point to our script and then within it call the run function as shown below:

```python
if __name__ == '__main__':
    run()
```

Note, run is a function because it's declared outside of the
BMICalculator class.

Access the full program here: http://bit.ly/33dIs8Q

Project: The Secret Number Game

We want to create a classic number guessing game in
which the player in essence plays against the computer.
Given a range of possible numbers from 1-100 we can
create a game. Here's some screenshots from that game:

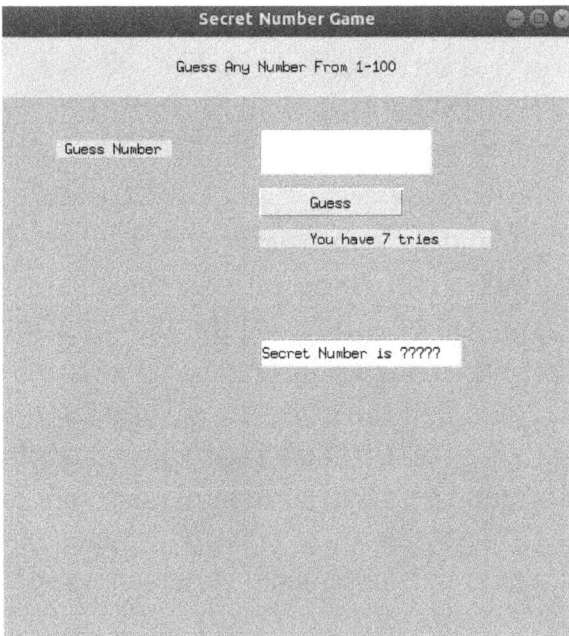

Image: Guessing Game 1.

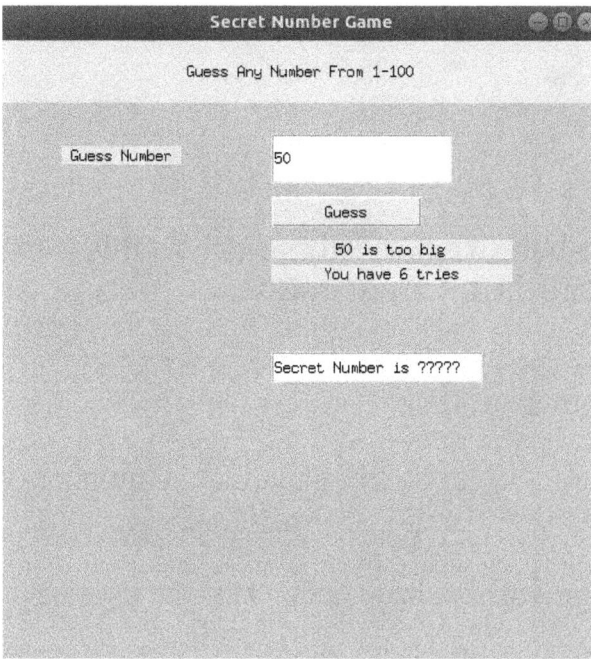

Image: Guessing Game 2.

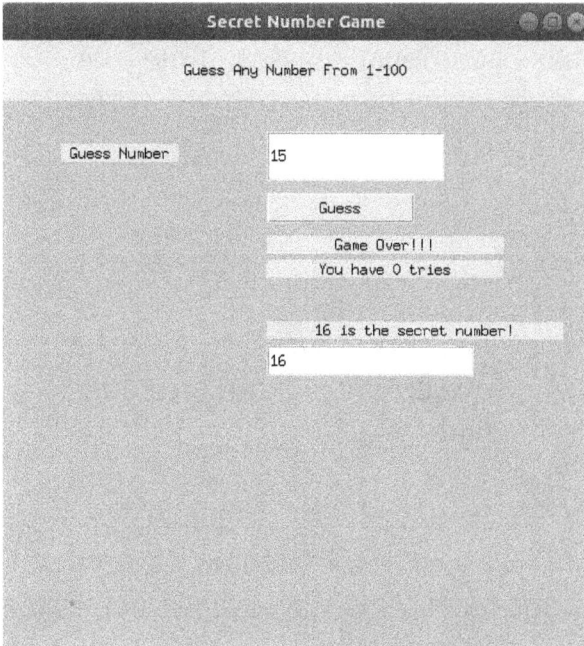

Image: Guessing Game 3.

The above screenshots in a nutshell shows how this game works. But, here are the rules just in case:

1. The program will randomly generate a secret number from 1...100.

2. The program will randomly generate the number of tries that the user can take form 1... 15.

3. The program will list how many tries the user has left.

4. If the user ran out of number of tries then the program will notify them of that and then display a message that the game is over. The user should not be able to make additional guesses. If the user guesses the right secret number then the correct number will be displayed.

Script Hint

Like the previous GUIs we created we can break this program down into methods.

- __init__: This is where we'll initialize the variables that will be used throughout the program.
- create_widgets: This is where the Label, Entry, and Button widgets are created.

- **guess**: This is where the logic is done that determines if the user input is equal to that of the secret number. A loop is actually not needed because an <u>implicit infinite loop is done</u> when the **mainloop** method is called.
- **run**: This does all of the setup for the GUI.
- **if __name__ == '__main__'**: This serves as the entry point to the program.

Solution

Now, let's analyze the code…

```python
from random import randint
from tkinter import Tk, Label, Entry, Button, StringVar
class SecretNumberGame:
    def __init__(self):
        self.root = Tk()
        self.secret_number = randint(1, 100)
        self.tries = (randint(1, 15))
        # a list of StringVars
        self.secret_num_str_var = StringVar()
        self.secret_num_str_var.set('Secret Number is
?????')
        self.your_guess = StringVar()
        self.hint = StringVar()
```

All of the variables that are used throughout the program will be added here. The next step is to create a method that setup all of the widgets that will be used in the program:

```python
def create_widgets(self):
    # Labels
    game_banner = Label(text='Guess Any Number From 1-
100').place(x=0, y=0, width=500, height=50)
```

```
    guess_number_label = Label(text='Guess
Number').place(x=50, y=85, width=100)
    your_tries = Label(text=f'You have {self.tries}
tries').place(x=225, y=160, width=200)
    # Entry Widgets
    guess_number_widget =
Entry(textvariable=self.your_guess).place(x=225, y=75,
height=40, width=150)
    the_secret_number =
Entry(textvariable=self.secret_num_str_var).place(x=225,
y=250, width=175,

height=25)
    # Button
    guess_budget = Button(text='Guess',
command=self.guess).place(x=225, y=125, width=125)
```

Three labels, two entries, and one button widget are created here. The place method is again used to layout the widgets in the GUI. A header is created that serves as a way to add a little style to the GUI. Next, let's dig into the **guess** method which serves as the logic to the game play. Below is the guess method:

```
def guess(self):
    """
    method updates the GUI
    after each user guess.
    """
    your_guess = self.your_guess.get()
    your_guess = int(your_guess)
    if self.tries == 1:
        your_tries = Label(text=f'Game
Over!!!').place(x=225, y=160, width=200)
        self.secret_num_str_var.set(self.secret_number)
        Label(text=f'{  self.secret_number} is the
secret number!').place(x=225, y=230, width=250)
        your_tries = Label(text=f'You have {0}
tries').place(x=225, y=180, width=200)
    elif your_guess > self.secret_number:
        self.tries -= 1
```

```
        Label(text=f'{your_guess} is too
big').place(x=225, y=160, width=200)
        your_tries = Label(text=f'You have {self.tries}
tries').place(x=225, y=180, width=200)
    elif your_guess < self.secret_number:
        self.tries -= 1
        Label(text=f'{your_guess} is too
small').place(x=225, y=160, width=200)
        your_tries = Label(text=f'You have {self.tries}
tries').place(x=225, y=180, width=200)
    else:
        self.tries = 1
        your_tries = Label(text=f'You
Win!!!').place(x=225, y=160, width=200)
        self.secret_num_str_var.set(self.secret_number)
        Label(text=f'{self.secret_number} is the secret
number!').place(x=225, y=230, width=250)
```

The first thing we need to do is get the user input which is done in this line of code:

```
your_guess = self.your_guess.get()
```

However, since the input is received a **StringVar** will be converted into an **int** which is done in this line snippet:

```
your_guess = int(your_guess)
```

Easy enough ok, so as mentioned in the hint we don't have to include a loop because an infinite loop is running in the background once the script starts. However, what we need to do is start with a counter and decrement the counter each time until we've found a match or until the counter is equal to 1. In our case the counter is **self.tries** which is the random number of tries that's generated within the range of 1… 15. Here's the code snippet for this:

```
if self.tries == 1:
        your_tries = Label(text=f'Game
Over!!!').place(x=225, y=160, width=200)
        self.secret_num_str_var.set(self.secret_number)
        Label(text=f'{  self.secret_number} is the
secret number!').place(x=225, y=230, width=250)
        your_tries = Label(text=f'You have {0}
tries').place(x=225, y=180, width=200)
```

If `self.tries` is equal to 1 then the `Label` is updated to let the player know that the game is over.

The remaining parts of the chained conditional checks to see if the player's guess was larger or smaller than that of the secret number. If those conditional checks evaluate to `False` then we can assume that the user guessed the correct number. If that's the case then the Label is updated letting them know that they guessed the correct number. The `self.tries` instance variable is set equal to 1 so that the user can't continue guessing after they won because that's not acceptable game play. The run method simply sets up the GUI, calls `create_widgets`, and then calls `mainloop`:

```
def run(self):
    """
    Setups the basics of the app.
    """
    self.root.title('Secret Number Game')
    self.root.config(bg='tan')
    self.root.geometry('500x500')
    self.create_widgets()
    self.root.mainloop()
```

To start the GUI use the if __name__ == '__main__' condition:

```
if __name__ == '__main__':
    SecretNumberGame().run()
```

And that's all there is to the Secret Number Game GUI.
You can view the full source code on GitHub:
http://bit.ly/323N3sW

Chapter V Notes

The next three pages are for jotting down any notes about chapter V.

Chapter V Notes

Chapter V Notes

Chapter V Notes

Book Resources

To view the urls in this book online you can access url:
http://bit.ly/2qYXH7E

To view this book repo on GitHub visit this url:
http://bit.ly/2N2d9bJ

INDEX